P9-CKY-705

Chemicals from Petroleum

The world of petroleum chemicals—a striking vista from near Houston, Texas

Courtesy Petro-Tex Chemical Corporation

Chemicals from Petroleum

An Introductory Survey

A. Lawrence Waddams ARCS BSc DIC MIChemE

THIRD EDITION

A HALSTED PRESS BOOK

John Wiley & Sons New York

First edition 1962
Reprinted 1963
Second edition 1968
Third edition 1973

Published in the U.S.A. by
Halsted Press, a Division of
John Wiley & Sons, Inc., New York

Library of Congress Cataloging in Publication Data

Waddams, Austen Lawrence.
Chemicals from petroleum; an introductory survey

"A Halsted Press book."

1. Petroleum chemicals. I. Title.
TP692.3.W3 1973 661′.804 73–3397

ISBN 0–470–91303–7

Printed in Great Britain

Contents

Illustrations

PLATES

DIAGRAMS

Sources of Flow Diagrams

The flow diagrams in this book are all taken, by permission of Gulf Publishing Company, from *Hydrocarbon Processing*, November 1971—© Gulf Publishing Company, Houston, Texas, 1971. Full details of sources are as follows:

page 25 'Ethylene' —C. F. Braun and Co.

page 66 'n-Paraffins (IsoSiv process—kerosene range)' —Materials Systems Div., Union Carbide Corp.

page 83 'Ethylene oxide'—Scientific Design Co. Inc.

page 122 'Isopropanol'—Tokuyama Soda Co. Ltd.

page 166 'Isoprene (USSR process)'—The Power Gas Corp. Ltd.

page 182 'Ammonia'—The M. W. Kellogg Co.

page 198 'Methanol (ICI low pressure process)'—I.C.I.

page 229 'Phthalic anhydride'—B.A.S.F.

page 244 'Caprolactam (low sulphate process—DSM)'—Stamicarbon NV.

Preface to the Third Edition

The prefaces to earlier editions indicated that one of the intentions of this book was to act as an up-to-date picture of an important segment of the chemical industry, for students and members of the academic professions. The book has also been used as an introduction to the scope and functioning of world petroleum chemical activities for people already within the chemical process industries.

As it is an industrial scene that is being described, the terminology used is that of the industry—with explanatory notes where necessary.

The history of petroleum chemicals is short but subject to rapid reassessment. The interval since the previous edition of this book is only four years, and yet some three-quarters of the earlier text required to be completely rewritten.

During the 1960's the emphasis in petroleum chemicals was on growth and innovation. These factors remain, but on a somewhat muted scale. The notable features of the present scene include a changing economic climate and a renewed influence of the petroleum industry on its chemical links. Two wholly new chapters have been included to discuss these topics.

Whilst it is not possible to point to a large number of entirely new processes which have reached commercial significance in the past three or four years, new items in the current edition include the production of isoprene from acetylene and acetone; a range of developments in the production of chlorinated hydrocarbon solvents from ethylene; new processes for the separation of xylene isomers; some developments in the plastics field including polybutene-1, nylon 12 and polybutylene terephthalate; disproportionation reactions involving olefins and aromatics; potential new outlets for ethylene in synthetic pulp and ethylene alkylate; developments in maleic anhydride production from butylenes; and a number of modifications in polyolefin processes. A whole host of processes have been modified and developed in detail.

As before, the opportunity has been taken of updating the figures, and in some instances adjusting the sequence to a more logical pattern. The outline story of the availability and production of the major raw materials, such as the lower olefins, aromatics and synthesis gas, has been rounded out further. This edition is some thirty per cent larger than the previous one.

It would have been difficult for the author to contemplate a new edition of this book without the consent and encouragement of the management of BP Chemicals International Ltd. Acknowledgement and thanks are due to this company for various forms of assistance which have been received. Likewise, colleagues in BP Chemicals have been most helpful on many occasions.

A number of flow diagrams, typical of the processes involved, have been included. For these, acknowledgement is due to the Gulf Publishing Co. of Houston. They are based on flow sheets appearing in the November 1971 issue of *Hydrocarbon Processing*. Full details of sources are given on page viii.

Finally, it is difficult to overestimate the contribution to this work of Miss Mollie Baker. She typed the entire text (including some telepathic moments when she was able to type what the author meant rather than merely what he wrote) and collaborated throughout in checking the typescript and the text.

April 1972 A.L.W.

Part I

Background to Petroleum Chemicals

Chapter I

Introduction

The chemical industry is of ancient origin, and is founded on a wide variety of sources of raw materials. Amongst the most important historically of these have been coal, molasses, fats and oils (of vegetable and animal origin), salt, metalliferous ores, water and the atmosphere.

The appearance of petroleum as a significant source of chemicals (if we exclude the special case of carbon black) dates back only to the middle 1920's. The early development of chemicals from petroleum, for reasons discussed in the next chapter, took place almost exclusively in the U.S.A. This development has continued in the U.S.A. with practically no check until the present day. It is still true to say that the petroleum chemical production of the U.S.A. is appreciably greater than that of the whole of western Europe—the next largest complex—though by 1980 the two should be comparable in size.

Petroleum chemicals form part of the chemical industry, so that their substantial development can best be anticipated in areas with a highly developed chemical industry. The early limitation restricting development beyond the U.S.A. was the lack of availability of raw materials. Prior to 1939 it was the policy of the petroleum industry to locate refining capacity in close proximity to the source of the crude oil. After 1945 (for a variety of reasons, beyond the scope of this book to explore) it became the policy to locate oil-refining capacity in the regions of major consumption. This resulted in a vast expansion in the refining capacity of western Europe, and caused attention to be diverted to the possibility of chemical production from these refinery resources. Since 1950 the development of petroleum chemical production in Europe has been an outstanding feature even in a picture of widespread expansion of industrial activity. The final chapter of this book provides the statistical picture of these developments.

Moreover, the use of petroleum as a basis for chemical production has now spread to new areas. The development of Japan's petroleum chemical industry in the 1960's surpassed even the European progress of the 1950's, which seemed spectacular enough at the time. Australia and Mexico are amongst the countries that have based some of their chemical production on petroleum raw materials for a number of years. Countries of Asia, South America and the Middle East now have a substantial number of plants in operation and others in the planning stages. Canada was early in the petroleum chemical field and has found the vast industrial potential of the United States both a stimulus and a limitation to her own development. Efforts are being made to put the considerable oil and gas resources of North Africa to chemical use, but these are at a fairly early stage. In South Africa there is a competitive focus for organic chemical developments in the great coal chemical complex at Sasolburg.

In summary, however, the industry which is responsible for producing chemicals from petroleum is one of worldwide significance.

Since petroleum is essentially a mixture of hydrocarbons, the chemicals made from it are nearly all organic chemicals. They stray over into the inorganic field only for specific reasons. Carbon black and hydrogen cyanide are commonly classed as inorganic on an arbitrary basis. Sulphur is present as an undesirable impurity in certain crude oils and in natural gas, and may be recovered either as the element or as sulphuric acid. In ammonia production, it is the hydrogen which is required from petroleum. There is another special influence here, in that the carbon content of the petroleum may, in effect, be replaced by hydrogen by application of the shift reaction as outlined on page 188. These examples outline the limited range of inorganic chemicals which may commonly be derived from a petroleum source.

In theory virtually any organic chemical may be made from petroleum. In practice, whilst the range is wide, it is not unlimited. Other raw materials have a well-entrenched place in the organic chemical field, and in a few cases such production is still expanding. Fermentation chemicals have declined in most areas. The issue is generally one of economics. In many cases the same compound can be derived from more than one raw material. There will obviously be a tendency to use the most economic raw material. Such a trend is likely to be gradual, firstly because relative economic conditions will vary between one area and another, and secondly because production in an older plant has the advantage of lower fixed charges (for two reasons: the normal effect of inflation, and the effect of plant depreciation allowances over a period of years). Moreover, there is a general rise in chemical production as a whole, so that even where a raw material may be losing ground relative to its competitors, it may still retain its position, or even improve it, in terms of absolute tonnage.

Already it seems evident that petroleum must play an increasingly predominant part in the production of organic chemicals. Probably nearly 90 per cent of the world's organic chemicals are already derived from petroleum hydrocarbons and it is commonly expected that all other raw materials will be responsible for no more than 1–2 per cent of organic chemical production by the end of this century.

Some of the original stimulus behind the development of chemical derivatives from petroleum was the appearance of products which, from an early stage, lent themselves particularly to this type of manufacturing operation (one may cite polyethylene as an example of this) and which then proceeded to develop into huge commodities in their own right.

Fermentation is at this stage generally applied only to the production of specialized groups of chemicals for which no simple synthesis has yet been discovered.

The European coal-chemical industry was at one time a vast and predominant producer more particularly of aromatics and acetylene. Most European benzene was made from coal up till the middle 1960's. The operation of coal carbonization, which is the main source of benzene from coal, proved insufficiently flexible and inadequate in scale to meet the demands for benzene in the later 1960's, particularly as the use of coal for making town gas was sharply reduced. Benzene is now obtained directly from refinery catalytic reformers and by extraction from the gasoline stream obtained when naphtha is cracked to ethylene. In addition benzene may be obtained indirectly from the hydrodealkylation of refinery toluene.

Acetylene as a raw material has faced acute competition on many fronts in recent years, and in any case most new capacity for acetylene production has been based on the high temperature pyrolysis of various hydrocarbons in the form of refinery products or natural gas.

The availability of a range of petroleum raw materials opened up a number of new horizons for the chemical industry. Some were economic, some chemical and some technological. The successful development of the opportunities thus created was a challenge to both the petroleum and the chemical industries. This book records how that challenge has been met.

Chapter 2

Characteristics of Petroleum Chemical Manufacture

Petroleum chemical manufacture represents that segment of the chemical industry which is concerned to produce chemical products from raw materials of petroleum origin. Operating in this field are petroleum companies who have extended their interests into chemicals, chemical companies who buy in their petroleum raw materials, and joint ventures between chemical and petroleum companies, amongst many others (a very mixed bag, including plastics producers, textile companies, shipping companies and a sprinkling of governments). The manufacture of petroleum chemicals has therefore proved a meeting ground for the petroleum and chemical industries, introducing some features novel to each, and creating an entity with a character somewhat distinct from either of its parent industries.

Such differences should not be laboured, however, for there is no doubt that the rightful place for petroleum chemical production is solidly within the confines of the chemical industry.

The petroleum industry has its own cherished traditions. Above all else, it is big. A medium-sized refinery is likely to consume five to ten million tons of crude oil each year. The capacity of individual plant units is, therefore, huge by the standards of most industries, but will represent quite a modest proportion of the total market for the appropriate products in the market area serviced.

Almost all petroleum products can be handled in a fluid form. Refineries, therefore, present a spectacular array of pipe tracks and tank farms, for the handling of raw material and finished products. Petroleum technology is favourably influenced by this fact. Such advanced techniques, for example, as the fluid catalyst bed (in which control of the catalyst movement is exerted by the flow of the reactants) became established some years ago. The whole background, in short, favours the development of large-scale, continuous, highly automated operations. High temperatures and catalytic promotion are frequently resorted to, but extremely high pressures and severe conditions of corrosion stress (such as are found in certain chemical manufactures) are uncommon.

Petroleum products are very rarely identifiable chemical compounds. They are more commonly prepared as blended products, required to meet a series of physical specifications, with occasional chemical interpolations as to impurities. This factor provides oil-refining operations with a significant degree of flexibility. Many of these blended products are prepared for markets whose standards of acceptance are well known and of long standing.

The phenomenon of the 'performance specification' is not unknown in the petroleum industry, but performance specifications for oil products are more commonly amenable to specific techniques of assessment operated within the refinery area (for example, the octane number of gasoline).

The chemical industry is rather more diverse. It ranges all the way from large-scale plants (still pretty small by petroleum standards, however), to small untidy batch-operated units, where, it appears, the spirit of alchemy sometimes still breathes. The chemical technologist has to face raw materials which include coal, salt, and metallic ores. The products are commonly pure compounds, with specifications governed by closely defined standards of chemical purity.

Petroleum chemical manufacture commonly requires the application of typically petroleum processing techniques to typically chemical finished products. Both the oil and the chemical partners in such enterprises found they had something to learn.

Petroleum chemical units are usually continuous, elaborate, operating with catalytic promotion, and highly automated. They consequently require a large scale of operation to secure an economic advantage. Unless there is an economic advantage, there is little purpose in the development of petroleum raw materials to serve the chemical industry in any particular instance. By their nature, petroleum chemical units are very much subject to the consideration of 'minimum economic size', and it is necessary to digress for a moment to establish this point.

The question of plant size assumes economic significance in any operation where the capital cost of the plant represents, in the form of depreciation, maintenance, etc., a substantial factor in the cost of the finished product. This situation is almost universally encountered in petroleum refining and petroleum chemical manufacture, but is slightly less common in some sections of the traditional chemical industry, where raw material costs may be of overwhelming importance. The economic importance of determining the most suitable plant capacity derives from the fact that the cost of building a plant is not proportional, in a simple sense, to the capacity (in practice, it is frequently assumed that the capital cost of plant will vary in proportion to the capacity taken to the power 0.6). The net effect of this, expressed simply, is that building a plant of double capacity will not require double the cost. So long as the plant can be operated at its maximum capacity, it is clear that there is an economic incentive to design it on the largest possible scale. At the lower end of the capacity scale, the effect of depreciation increases by leaps and bounds. In a given set of conditions, therefore, which will reflect the facilities operated by competitive companies, it is important to ensure that any project envisaged includes production facilities of not less than 'minimum economic size'.

The past few years have caused a startling revision in the concept of what constitutes an economic size of plant. Spearheaded by the large tonnage basic chemicals, such as ethylene and ammonia, the scale of individual processing plants has increased to a spectacular extent. The typical ethylene plant in Europe in 1956 would have a capacity of 30 000 tons per year; in 1962 this would have risen to 70 000 tons per year, and now to plan a unit of less than 100 000 tons per year is almost unthinkable. Normal capacities are in the range of about 250 000 to 400 000 tons per year. The top limit is creeping up to rather beyond 500 000. For many years this enabled the industry to make the happy assumption that a combination of new products, process innovation and the economic benefits of increasing scale would allow continuously falling prices in an inflationary world.

In practical terms there are economic limits to the scale of most manufacturing operations. These may be associated with the mechanical aspects of engineering. Beyond certain sizes some items of equipment are unlikely to be available (and duplication of single items destroys the whole basis of obtaining a benefit from scale). Other items become virtually incapable of transportation as a single unit and may therefore require some degree of site fabrication with attendant cost disadvantages. Above all, with increasing size of plant the economic benefits of increasing the scale still further become less evident, while the risks occasioned by a late start-up or a malfunctioning of the plant increase at least arithmetically with scale. A further factor, which one may hopefully postulate as temporary, has been the great acceleration in capital costs of building such plants in the past three years.

These brief comments will serve to indicate that at last inflation seems to have caught up with petroleum chemicals and that any further benefits of scale will be relatively modest.

Although we find, at a given moment, practical maximum limits of size that are set by current engineering capability, a more common requirement is still to establish the minimum economic size of plant that can be viable. This must, of course, relate to a particular operation in a particular place. Having established the minimum size, it is necessary to take a clear, cold look at the market available. A single petroleum chemical unit may be expected to serve a much higher proportion of the total market than most petroleum units would do. Since the final product is likely to be the subject of precise chemical specification, there is little flexibility in production available to such units. A new chemical product has to find its market either in competition with an accepted product, or in an entirely new field. The building of a plant of too great a capacity, therefore, will leave it under-occupied for an extended period. This factor may be accentuated, in the case of a new product, by the necessity to secure its full technical acceptance before market development can proceed.

The increasing movement of petroleum chemicals into the plastics field introduces some further novel and troublesome elements. The properties of plastics are rarely defined adequately in terms of chemical composition and such physical properties as melting point, viscosity, etc. Such items may well form part of the specification, but the ultimate requirement can only be expressed in terms of performance. Here, however, the 'performance specification' is a good deal more difficult to establish than is common in the field of petroleum. The customer requires the plastics he purchases to perform to his satisfaction, in his equipment, on his premises. The fabrication of plastics involves a multitude of small enterprises, calling forth much individual initiative and catering for a variety of specific individual requirements. Whilst this situation has permitted the dynamic growth of the industry as a whole, it has also created the need for a very high standard of specialized technical service. This background of experience was entirely new to those brought up in the traditions of petroleum.

Just as the oil partners in the petroleum chemical complex had something to learn, the chemical partners have also secured some benefit from the association with techniques borrowed from the oil industry. One factor may be illustrated by reference to manpower statistics. A series of figures quoted by the O.E.C.D. may be typified by the calculation that chemical production in the O.E.C.D. countries of western Europe rose 80 per cent between 1963 and 1968 whereas employment in the industry rose a bare 8 per cent. In passing it may be mentioned that in the U.S.A. during the same period an increase in chemical production of 50 per cent was accompanied by a rise in employment of 20 per cent. Before Europeans take too conceited a view of their performance it should be noted that in 1963 the U.S. productivity (in terms of added value of production per person employed) was almost exactly three times the European average.

Clearly a major factor in the improved productivity has been the continuous growth of petroleum chemical production with its emphasis on automation and relatively low manpower requirements. Certain elements in the chemical industry do not, of course, lend themselves to large-scale continuous automated production units. Fine chemicals, for instance, are generally produced in diverse, small-scale, versatile, batch-operated units.

It was not unreasonable, when a traditional chemical plant was behaving oddly, to shut it down, open it up and find out the reason. With the large-scale capital-intensive petroleum chemical plants, a shutdown is a costly and disruptive affair, to be avoided if at all possible. The level of maintenance effort required on these plants is exceptionally high, usually well in excess of the effort (expressed in terms of numbers of men) required for plant operation. These were among the lessons to be learned by the chemical companies operating in this field.

The specialized nature of the processes and requirements of the petroleum industry—including the occasional necessity to build a refinery on some inhospitable rock, thousands of kilometres from the resources of industry—has created the specialist contractor to deal with this type of design and construction. The emphasis in such contract work is on detailed planning, well ahead of the proposed construction schedule. The contractors represent a concentration of facilities and expertise from which chemical and other companies can draw. The fluctuations in the construction programme of a specific company are such that it may be inconvenient and uneconomic to maintain a permanent construction organization. The facilities of contractors represent a pool from which individual companies may draw according to their specific needs. The arguments for calling in a major contractor in any particular instance are clearly a matter for individual judgement, but there is little doubt that the chemical industry has benefited to some extent by an acquaintance with the services that the expert contractor can offer.

One advantage that petroleum chemical manufacture derives from the background philosophy of the oil industry is the traditional freedom with which processes are made available within the industry by licence and royalty agreements. The policy of the traditional chemical industry has been rather more cautious in this respect.

Perhaps the greatest mark of esteem which a process in the classical tradition of chemical synthesis could acquire was the description 'elegant'. Such a description is commonly applied to an indirect synthesis, involving several stages, each of which gives a high yield of product, and commonly including some novel or unexpected feature. The new technology is a little averse to this approach. The tendency now is to seek out the simplest form which a synthesis can be theoretically made to take, and then to try and make it work in practice. All the resources of pressure, temperature and catalytic promotion are brought into play with the aid of the most highly developed materials of construction. Any steps in the synthesis which can conceivably be avoided are cut out. All raw materials, other than those represented in the final product, are eliminated as far as possible. In developing this approach it is necessary to accept relatively low conversions, and consequently considerable recycling of reactants. Processes derived in this manner may not, in the classical sense, always be termed 'elegant', but they tend to achieve the maximum degree of economy. The capital costs of these new plants are generally high, but the operating costs are kept down, so that full benefit can be reaped from operating the process on a large scale. Examples illustrating some aspects of this technological approach include the direct hydration processes for the manufacture of ethyl and isopropyl alcohols, the direct oxidation processes to ethylene oxide and acetic acid, the acrolein route to glycerine, and the propylene route to acrylonitrile.

The location of petroleum chemical facilities has until quite recently been subject to a number of very limiting factors. The U.S.A. had a good start before 1939, because it was alone in possessing raw material resources (in the form of oil fields, natural gas fields and refineries) together with a substantial market for the chemical products. It is not always properly understood that the chemical industry thrives best in an industrial climate. Only in such a climate will exist the major consumers of its products, which the large-scale operations of petroleum chemical manufacture demand. Moreover the production of chemicals from petroleum involves a considerable consumption of such chemicals as chlorine, sulphuric acid, etc. These are generally available on an economic basis only in an industrial country. The building of refineries in Europe since 1945, and the discovery of substantial natural gas resources, formed the basis for the extension of petroleum chemical production to the European continent. This made quite a slow start, for the chemical industry based primarily upon coal, was already well entrenched. The expansion in Europe since 1955 has been very rapid, and some indication of this will be found in the final chapter.

The above outline will indicate the difficulties of initiating petroleum chemical manufacture in less industrialized areas, even where the raw materials are cheaply and abundantly available. One has been faced with the alternatives of building a plant merely to meet local needs, in which case it will be well below the normal economic size and likely to require a government subsidy or massive protection to be viable, or of building the plant to an economic size and then to hunt for external markets. Until very recently this has posed considerable problems. Chemical plants tend to be more expensive to construct in a developing country. Most of the adjacent markets are likely to be relatively small, and the expertise of industrial exporting will be lacking. In the past few years there have been signs that these difficulties can be overcome in certain instances. The techniques of bulk transport of basic chemicals are developing to the point at which most of the production from a large plant sited in an area of cheap raw material can be economically shipped in bulk to consuming areas. Some of the early plants being established in the Middle Eastern areas are making bulk products such as ammonia and sulphur which lend themselves to these modern transportation techniques. In the future, ethylene must be regarded as a possibility.

The manufacture of bulk petrochemicals in large units may therefore be expected to develop to an increasing scale in oil-producing areas. In spite of such trends, it may still be observed in general that petroleum chemical production has developed fastest and furthest in areas with a high degree of industrialization, including a well-developed chemical industry.

It is relevant, in this chapter devoted to characteristics of petroleum chemical manufacture, to refer briefly to some of the techniques which are used, and to which reference is made in subsequent chapters. Most of these are borrowed from petroleum technology, and several will be familiar in a more general sense. Amongst the processes of importance in petroleum chemical manufacture are:

(a) *Distillation*

Distillation is a technique of separation which makes use of the difference in volatility or boiling point of different components in a mixture. It is effected in a column containing a series of horizontal perforated plates or trays spaced inside. The aim is to carry the most volatile component of the mixture away in the vapour phase from the top of the column, and to draw off the least volatile component as a liquid from the bottom (alternatively, according to specific requirements, a side stream may be taken from almost any point up the column). To effect complex or difficult separations, it is probable that more than one column will be required.

To achieve close control of column operation and a high purity of the products it is normally appropriate, having condensed the overhead vapours from a distillation column, to return a proportion of the condensed liquid to the top of the column as reflux. The operation then becomes known as fractional distillation or fractionation.

The term distillation is properly applied only to those operations where vaporization of the liquid mixture yields a vapour phase containing more than one constituent. Where the vapour from a liquid mixture or solution (such as brine, or a glycerine and water mixture containing less than 80 per cent of glycerine) contains only one component, the process is described as evaporation.

Distillation may proceed at atmospheric pressure, under high vacuum, or at elevated pressure, in accordance with the specific requirements of the separation. The industrial significance must be stressed of distillation at very low temperatures. A combination of refrigeration, and operation at moderate pressure, enables the distillation to be effected of components which would be gases in normal circumstances.

In an endeavour to assist the separation effect, it is sometimes necessary to modify the process of fractional distillation, e.g. by using additional components. This results in such operations as steam distillation, extractive distillation (distillation in the presence of a solvent which becomes the least volatile component of a mixture), and azeotropic distillation (distillation in the presence of a solvent which becomes the most volatile component of a mixture).

(b) *Solvent Extraction*

This is the separation of a component or components of a mixture by the use of a liquid with selective solvent characteristics. This operation is used for the separation of components by types: for example, the separation of aromatics from paraffins.

(c) *Crystallization*

By allowing a component to form crystals from solution, or from a molten mixture, the solid crystals can be separated by filtration or centrifugal separation from the other components.

(d) *Absorption*

This process is a form of solvent extraction. A component of a gas or vaporized mixture is separated by selective absorption, usually in a liquid solvent. The operation is commonly carried out in a packed tower.

(e) *Adsorption*

Certain highly porous materials (e.g. activated charcoal, silica gel, molecular sieves) have the power of condensing on their surfaces large amounts of vapours. Where this adsorption can be operated selectively, it represents a technique for the separation of one component from a mixture.

(f) *Cracking*

In petroleum terminology, this means the breaking down of the large hydrocarbon molecules into molecules of lower molecular weight.

This is achieved in the absence of air, by high temperature alone (in which case it is called thermal cracking or pyrolysis) or by a combination of high temperature and catalytic activity.

Amongst the low molecular weight products of cracking are the lower paraffins and olefins, which represent major raw materials for chemical production.

Thermal cracking at low pressure in the presence of substantial qualities of steam is a modification normally designed to provide high yields of unsaturated gases.

(g) *Reforming*

This is a term normally used in petroleum technology to refer to processes designed to upgrade gasoline quality. Straight run gasoline is subjected to thermal or catalytic reforming to modify the molecular structure of its components, so that the final product has a higher anti-knock rating. Reforming is now more commonly carried out catalytically and in the presence of hydrogen.

The term reforming is also used more loosely in connection with a variety of hydrocarbon reactions (e.g. the methane-steam reaction) which will form synthesis gas or town gas.

(h) *Alkylation*

This term is used to describe a reaction in which a straight-chain or branched-chain hydrocarbon group (called the alkyl group) is united either with an aromatic molecule, or an unsaturated hydrocarbon group (which may be straight-chain or branched-chain), to form a new complex molecule. In the chemical industry the application of this process is most common in the production of alkylated aromatic compounds. In the oil industry a similar process is of considerable importance in the production of high-octane components of gasoline (iso-octane itself being a typical product).

(i) *Isomerization*

This is a process designed to induce a rearrangement of atoms within a particular molecule. In the petroleum field it is commonly applied to the conversion of a normal paraffin to the isoparaffin. Isomerization of methyl cyclopentane to cyclohexane is of significance in the application of catalytic reforming techniques to aromatics production.

(j) *Polymerization*

No simple definition of polymerization will satisfy all the ramifications of this complex operation. In essence it is a chemical reaction in which the molecules of the single reactant are linked together to form large molecules whose molecular weight is a multiple of that of the original substance. The initial reactant is called the monomer and the final product the polymer. Where two or more monomers are involved the process is called copolymerization. When the polymerization is restricted to a specific stage of development, a combination of two molecules of monomer is called a dimer; with three molecules in combination it becomes a trimer, and so on.

The following table provides indications of the properties of various petroleum fractions:

Product	*Boiling Range*	*Hydrocarbon Molecule*
Gas and liquefied gas	up to 25 °C	C_1–C_4
Gasoline (petrol, naphtha)	ca. 20–200 °C	C_4–C_{12}
Kerosene	ca. 175–275 °C	C_9–C_{16}
Gas oil and Diesel oil	ca. 200–400 °C	C_{15}–C_{25}
Lubricating oil	—	C_{20}–C_{70}
Fuel oil	—	C_{10} upwards
Bitumen and coke	—	large molecules

The terminology for the hydrocarbon molecule is expressed in the normal shorthand form. The suffix represents the number of carbon atoms in the relevant molecules. It will be appreciated that these definitions are broad, and that some overlapping in fractions takes place.

Chapter 3

Raw Materials

Introduction

The original basis for petroleum chemical manufacture, as we now understand it, was the availability at refinery locations of olefin-rich gas streams as a result of refinery cracking operations.

Previously to this, natural gas had been used for the production of carbon black, but the more widespread use of natural gas as a chemical raw material, and in particular, its development to the manufacture of other products such as ammonia, followed closely after the earliest units based on refinery gases.

These two sources still represent a high proportion of the raw materials used for petroleum chemicals in the U.S.A.

The position may be illustrated by the basic product ethylene. The raw material sources for ethylene in the U.S.A. in 1970 comprised natural gas (ethane and propane) 75 per cent of the total, refinery gases 15 per cent, and liquid feeds 10 per cent. It has been suggested that the proportion of liquid feeds could rise to 50 per cent in 1980 and 75 per cent in 1990. Such forecasts tend to assume that appropriate quantities of naphtha can be made available from eastern hemisphere sources on a favourable basis. Such assumptions are open to question—more particularly as a vast new market for naphtha may be opening up in the production of 'synthetic natural gas' in the U.S.A.

The European pattern evolved somewhat differently. Both natural gas and refinery gases lack flexibility as raw materials. The production of refinery gases is geared to the requirements of petroleum products in both a qualitative and quantitative sense. The producer of chemicals from such a petroleum source can never, therefore, be sure that the availability of his raw materials will expand in line with his market for chemicals. In fact his experience in the past two decades has been the opposite. The petroleum industry itself is tending to become more and more specialized, and consequently requires an increasing volume of refinery gas for the preparation of special fuels and additives, whereas previously any gases not fed to the chemical plants were likely to be burnt in the refinery boilers. So far as Europe is concerned, therefore (with a larger chemical industry than the U.S.A. in relation to its petroleum and refinery resources), refinery gases represent a somewhat limited supply of raw material in terms of quantities available, more particularly when one is looking to future prospects.

Natural gas is used as a chemical raw material on a relatively modest scale in Europe. Supplies are not so abundant in Europe as in some areas of the U.S.A. Natural gas has also to face competitive pressure from the energy-producing industries (frequently operating under government auspices). A further vital consideration is that Europe's natural gas is mostly dry, consisting essentially of methane, with a rather limited range of applications as a chemical raw material.

In many areas of Europe, therefore, a third source of petroleum chemical raw material has been developed. This is represented by cracking units outside the refinery complex, though commonly connected to an adjacent refinery by pipelines conveying liquid feedstock to the cracker, and returning unwanted hydrocarbons to the refinery system. These cracking units are specifically designed to produce high proportions of olefinic gas streams from liquid hydrocarbon fractions in the range of naphtha to gas oil. They bear some superficial similarities to the older petroleum thermal cracking plants. The latter, however, were based on heavier feedstocks, and used elevated pressure and lower temperature as a means of minimizing gas formation since gasoline was the product sought. The naphtha cracking plants, on the contrary, are designed to provide a maximum gas yield, and so operate at effectively reduced pressure and high temperature.

In western Europe at the present time marginally over 90 per cent of ethylene is made from liquid feedstocks. Of the remaining production a small and declining proportion comes from refinery gas, and the remainder from natural gas. The only prospect of a material change in this position would be in the discovery of really substantial quantities of 'wet' natural gas in Europe—in the North Sea for example.

Some indications of the technological variations induced by the different raw materials may be given by the yield of lower olefins obtained. From ethane cracking a yield of 80 per cent wt. ethylene may be obtained. The remainder of the ethane is converted to fuel gas (methane and hydrogen) 12–13 per cent wt. and only 7–8 per cent is converted to fractions heavier than C_2. The yield of propylene is therefore very small.

From propane cracking a yield of 43 per cent wt. ethylene and 14 per cent wt. propylene is normal. The remaining propane is converted to fuel gas (ethane, methane and hydrogen) 27 per cent wt. and C_4 and heavier fractions 16 per cent.

These figures may be contrasted with the yield patterns from naphtha discussed later in this chapter.

Classification of Raw Materials

From the outline given, it is apparent that the basic raw materials for chemical manufacture are natural gas, refinery gases and liquid hydrocarbon fractions. To these should be added, to a lesser extent, certain types of wax. The waxes have not yet been mentioned, and form the subject of a separate section at the end of this chapter.

From these basic petroleum raw materials are derived the secondary raw materials, which form the basis of much of the classification used in the second part of this book. Secondary raw materials and their derivation may be summarized as follows:

Acetylene	from cracking or partial oxidation, either of methane in natural gas, or of higher paraffins.
Methane	a major constituent of natural gas.
Higher Paraffins	ethane, propane and butane separated from refinery gas streams or natural gas. Other raw materials in this category are paraffinic naphthas, and n-paraffins of varying carbon chain length (see Chapter 6).
Ethylene	present in some refinery gas streams. Produced by pyrolysis of propane and ethane (from refinery streams or natural gas) or by thermal cracking of liquid hydrocarbons.
Propylene	from refinery gas streams or by thermal cracking of propane and liquid hydrocarbons.
C_4 Hydrocarbons	from refinery gas streams or by thermal cracking of liquid hydrocarbons.
Higher Olefins	from wax cracking, or n-paraffin dehydrogenation. Also made by synthesis from lower olefins.

The derivation of these secondary raw materials from the basic raw materials is the subject which this chapter treats in outline. Certain aspects of this subject, such as the detailed separation of individual components of C_4 hydrocarbon streams for specialized applications, are dealt with later in association with the derivatives.

The above list of raw materials is by no means comprehensive. Amongst the raw materials of commercial importance may also be included synthesis gas, aromatics and sulphur compounds. A section of the second part of this book is concerned with the study of such raw materials and their derivatives.

Refinery Gases

Sources of Refinery Gases

Refinery gases may range from hydrogen, at the lower or more volatile end, to hydrocarbons with four carbon atoms in the molecule at the less volatile end. Essentially they include, in addition to hydrogen, the olefins and paraffins within the range indicated: from methane to the butanes in the case of the paraffins, and from ethylene to the butylenes amongst the olefins. According to circumstances there will be additional components present, usually in small proportions, such as acetylene, certain dienes (notably butadiene), and impurities such as hydrogen sulphide and nitrogen. The hydrogen sulphide is an undesirable impurity in a variety of petroleum processes (also in fuel gas), and its removal is normally effected at an early stage.

Refinery gases originally used for petroleum chemical manufacture came from the now obsolete process of thermal cracking. Some processes involving a degree of thermal cracking (e.g. coking, viscosity breaking) continue to be carried out in refineries, but they do not represent a major source of refinery gases. Today the three main sources of refinery gases are the processes of crude oil distillation, catalytic cracking and catalytic reforming. A further process, already of considerable importance in the U.S.A. but not yet very significant in Europe, is a variant of catalytic cracking known as hydrocracking.

The distillation of crude oil produces a volatile fraction of paraffinic gases. Only a limited proportion of such volatile material is an acceptable component of gasoline, and the major proportion is separated at the distillation stage, a process known as stabilization of the gasoline. Methane, ethane, propane and butane are the main constituents of this gas fraction. The proportion of gas produced will vary very widely between one crude oil and another.

Catalytic cracking has largely taken the place of thermal cracking in the refinery since it will provide a more valuable range of products, and in particular, a higher quality gasoline. A modern 'fluid-bed' type of catalytic cracker (e.g. one where the control of movement of the catalyst is maintained by the flow of the gases and vapours) uses its catalyst in the form of a fine powder. This catalyst normally consisted of alumina and silica in conjunction. Typical yields of gases from this operation are in the table below. It will be noted that there is a significant proportion of olefins in this gas stream.

Typical Gas Yields from Catalytic Cracking

Feedstock: gas oil
Total gas yield 16.5% wt. of which

hydrogen is	0.1% wt.
methane is	1.3% wt.
ethylene is	0.6% wt.
ethane is	1.9% wt.
propylene is	2.7% wt.
propane is	2.3% wt.
butylenes are	4.0% wt.
butanes are	3.6% wt.

Recent modifications in catalytic cracking have aimed at increasing the yield of the gasoline fraction. These modifications have included a very widespread switch to cracking catalysts of the molecular sieve type (in practice sodium aluminosilicates with the sodium ions replaced by polyvalent metal cations by ion exchange), changes in process technique (e.g. 'riser' cracking) and hydrogen treating of the feedstocks. There are consequently some significant variations in the yields of useful olefins obtained from these plants. The essential aim of such processes, after all, is to provide petroleum products rather than chemical raw materials.

A further extension of the process is called hydrocracking. As mentioned above this is already used to a major extent in the U.S.A. and is of potential importance elsewhere. This process is a combination of catalytic cracking and hydrogenation, in which the catalyst may be sodium aluminosilicate with the sodium ion replaced by a heavy metal such as palladium or nickel, carried out at moderate temperature and elevated pressure. The gases produced are saturated and include significant quantities of propane and the butanes, especially isobutane.

Catalytic reforming is another process designed to improve the quality of gasoline production. The process is applied to a heavy gasoline fraction including a proportion of naphthenes. In the presence of a catalyst, which is normally platinum-based, the naphthenes will be isomerized and dehydrogenated to produce aromatic compounds. A typical reaction is as follows:

methyl cyclopentane (ring: H_2C, $CHCH_3$, CH_2, CH_2, H_2C) $\xrightarrow{\text{isomerization}}$ cyclohexane (ring: CH_2, H_2C, CH_2, H_2C, CH_2, CH_2)

$\downarrow$ dehydrogenation

benzene (ring: CH, HC, CH, HC, CH, CH) $+ 3H_2$

[Note: this is commonly abbreviated to (benzene ring symbol)]

Catalytic Reforming Reactions

A second type of reaction of increasing importance is the cyclization of paraffins, which also produces aromatics. This reaction is particularly stimulated by the newer platinum–rhenium catalysts which can operate at relatively low pressures.

The overall effect is to increase the quality of the gasoline as measured by its anti-knock rating. The reactions take place in a reducing atmosphere, so that, apart from hydrogen, the gas fraction contains only saturated paraffinic compounds (together with a proportion of hydrogen sulphide if sulphur is present in the feedstock). Typical gas yields obtained from this process are quoted in the table overleaf. The volatile fraction of this gas stream is rich in hydrogen and can be used as a raw material for ammonia production.

Typical Gas Yields from Catalytic Reforming

Feedstock: naphtha
Total gas yield 15.0% wt. of which
hydrogen is 2.3% wt.
methane is 1.5% wt.
ethane is 2.1% wt.
propane is 3.8% wt.
butanes are 5.3% wt.

The gas yield may vary from under 10 to over 20 per cent wt.

The liquid fraction is essentially a gasoline that has the advantage of containing no unsaturates which can lead to gum formation in contact with air. Furthermore the C_6–C_8 fraction of the reformate has become a major source of aromatics, especially toluene and the xylenes. This fraction may be alternatively used as a high-octane component of gasoline or as a raw material for the extraction of the various aromatics as pure compounds.

Application of Refinery Gases

In considering the feasibility of producing chemicals from refinery gases one must pay attention to the concept of 'minimum economic size' to which reference has already been made. A small refinery will rarely produce by itself enough of any particular gaseous raw material to justify the building of a chemical plant to use it.

In the case of a substantial refinery, once it is proposed to set up a petroleum or chemical plant to use one or more components of the gas stream, a complicated gas separation process requires to be installed. Once this gas separation process begins, it tends to become increasingly worthwhile to use other gases which are separated or concentrated at the same time. In this way can be developed an integrated complex of plants, designed to make use of each of the separated gas streams as far as possible.

A substantial section of the whole range of petroleum chemicals is that based on the lower olefins. Even in refineries with a large catalytic cracker there is frequently insufficient of these olefins (particularly ethylene) in the gas streams for a worthwhile chemical development without some form of supplementation.

The production of additional ethylene in a refinery is commonly provided by a separate thermal cracking or pyrolysis operation using ethane, propane, or both as the charge stock. Typical cracker gas compositions from ethane and propane pyrolysis are as follows:

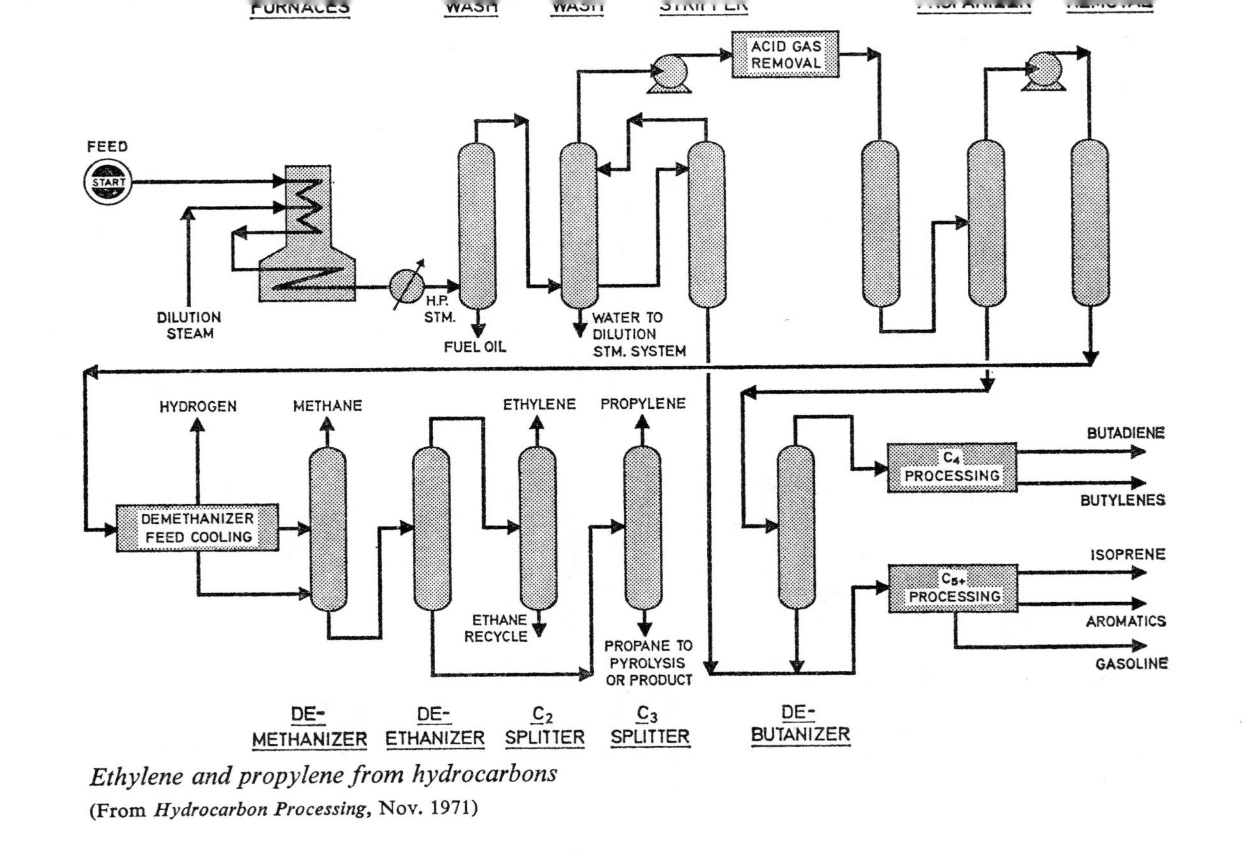

Ethylene and propylene from hydrocarbons

(From *Hydrocarbon Processing*, Nov. 1971)

Product in Cracker Gas	*Ethane Cracking % vol.*	*Propane Cracking % vol.*
H_2	36.7	16.1
CH_4	3.7	30.8
C_2H_2	0.2	0.3
C_2H_4	30.9	24.0
C_2H_6	27.1	3.9
C_3H_6	0.8	11.1
C_3H_8	0.6	11.3
C_4^+	—	2.5

The final stage of the operation for ethylene recovery is the C_2 splitter where ethylene is separated from ethane. Ethane is recycled to the pyrolysis unit where it produces additional ethylene, in accordance with the above table, as well as more hydrogen.

The C_3 fraction from the refinery gas separation unit (consisting of propylene and propane) may be fed to a propylene-consuming unit (such as an isopropyl alcohol plant). The propane passes unchanged through such a unit and may be fed to the pyrolysis operation to make more ethylene, propylene and, incidentally, hydrogen.

In theory it can easily be demonstrated that the more units assembled together in an integrated complex the more economic is the whole operation. In practice, it is not desirable to pursue this aim to the ultimate paper balance of raw materials and finished products, as the operation becomes so complicated that a minor fault in one of the units could create repercussions which could snowball into a major shutdown.

Natural Gas

Natural Gas as a Raw Material

Natural gas is a growing source of energy in the world, accounting at present for about 17 per cent of total world energy. It has long been an important source of raw material for chemical production in the U.S.A. Elsewhere it has assumed little commercial significance in this respect until very recent years. The trouble is that although abundant supplies of natural gas are available elsewhere, much of it is in countries with little industrial development, and consequently not offering large scope for the local use of petroleum chemicals. Such countries as Iraq and Venezuela have campaigned for the development of petroleum chemical manufacture based on their natural gas supply, and in the absence of a marked response from commercial interests, embarked on some ventures in this field, sponsored by the respective governments. Some similar developments, involving the participation of commercial interests, are now taking place in other oil-producing areas, such as Kuwait.

A development of greater commercial significance up to the present time has been the discovery of major natural gas fields in Italy and France. The more recent discoveries in the Netherlands and in the North Sea seem destined to be applied more actively to the production of various forms of energy rather than chemicals. Some of the French discoveries are of 'sour' gas containing substantial impurities of sulphur compounds. Until recently, this would have been regarded as a grave disadvantage, but in today's conditions it enables France to become one of the world's major producers and exporters of sulphur.

Natural gas available in western Europe, apart from such impurities as sulphur compounds and nitrogen, consists essentially of methane. This has so far obviated any possibility of developments according to the American pattern, whereby the higher paraffins present in natural gas are used as a major source of the lower olefins.

Where natural gas is virtually a methane stream, its range of chemical uses is limited but important. The major syntheses are by way of synthesis gas to ammonia, methyl alcohol, and Oxo alcohols, but there are significant applications for methane in the production of chlorinated methanes, hydrogen cyanide, and carbon disulphide. A further application is in the production of the secondary raw material acetylene.

Acetylene from Natural Gas

In considering natural gas as a raw material, it is necessary to take account not only of its major content of methane, but also of its potential for the production of acetylene.

The traditional route to acetylene is from calcium carbide, and hence indirectly from coal, according to the reactions:

$$CaO + 3C \rightarrow CaC_2 + CO$$
$$CaC_2 + H_2O \rightarrow CaO + C_2H_2$$

The alternative production of acetylene from hydrocarbons is feasible only where there is a cheap source of the hydrocarbon. The economics of this operation are usually assisted by the absence of cheap and abundant sources of local coal and power. It will be appreciated that whilst calcium carbide can be transported reasonably economically, once the acetylene has been produced, either from carbide or hydrocarbon, economical transport is rarely feasible. The required conditions first arose to a substantial degree in such areas as Germany and the Gulf Coast of the U.S.A. Currently about 54 per cent of the acetylene used for chemicals in the U.S.A. comes from hydrocarbons (largely natural gas). Outside the U.S.A. there have been developments along similar lines, notably in Italy and western Germany. The increasing availability of natural gas in western Europe offers a theoretical prospect of additional acetylene production from this source. The competitive future of acetylene has been in such doubt that there is little sign of such a development. Recent facilities for hydrocarbon-based acetylene production in Europe have almost all been designed to use liquid feedstock.

The reaction

$$2CH_4 \rightleftharpoons C_2H_2 + 3H_2$$

begins increasingly to favour acetylene production at temperatures in excess of 1 200 °C. There are three basic steps involved in such a process, each offering problems. First, it is necessary to introduce the heat required for the reaction to proceed; secondly the products of the reaction must be cooled very rapidly to prevent decomposition of the acetylene; and thirdly the acetylene must be separated from other reaction products.

One method of achieving the required energy input is by applying an electric arc to the feed. By this means its temperature is raised extremely rapidly to about 1 600 °C. This method has been operated for many years by Hüls in Germany. The rapid quench is now achieved by introduction of a cold secondary hydrocarbon feed, which not only cools the reaction products but provides a little extra acetylene. Yield of acetylene on total hydrocarbon feed is 34 per cent by weight.

A modified electric arc process has been announced (but not used) by du Pont. Electromagnetic forces are applied to stretch the arc in the direction of the gas flow and to cause it to rotate rapidly. Once again a secondary hydrocarbon feed is used as the initial quenching medium and it is claimed that the cracked gases will contain 22.5 per cent acetylene.

The Wulff process employs direct and regenerative heating for raising the feed to cracking temperature. This requires pairs of furnaces, each containing two masses of refractory material. In each furnace the heating run involves preheating air in one refractory mass (which is thereby cooled), raising the combustion chamber to cracking temperature using any appropriate fuel (including the feed itself) and allowing the combustion products to heat the second refractory mass. The cracking run operates in the reverse direction. Steam and feed are preheated by the hot refractory mass and pass through the cracking chamber; the cracked products are rapidly cooled by the second refractory mass. By operating the furnaces in pairs, the cycles can be made to yield a virtually continuous reaction. With methane cracking, the feed is raised to about 1 500 °C for a period of perhaps 0.03 second and the cracked gases contain perhaps 14 per cent acetylene. Acetylene separation is by means of the selective solvent dimethylformamide. This process languished at the pilot plant stage for many years. Several recent commercial units are based on a liquid hydrocarbon feedstock, and this operation is not without its difficulties. This process can be used to produce some ethylene as well as acetylene.

Probably the process in the most extensive scale of operation is the Sachsse or B.A.S.F.[1] process, which employs partial oxidation (i.e. controlled oxidation in the presence of a deficiency of the oxidizing medium to prevent the oxidation proceeding to its completion) as the means of providing the energy for the main reaction. The partial oxidation reaction is

$$CH_4 + 1\tfrac{1}{2}O_2 \longrightarrow CO + 2H_2O$$

and it is desirable to use oxygen rather than air to avoid unnecessary dilution of the gases to be heated. It is, therefore, necessary to incorporate an air separation unit in the equipment. A specially designed burner provides a reaction temperature of about 1 500 °C and the products are immediately quenched. Originally water quenching was used, but particularly with the extension of feedstocks used to light naphtha, it has been found more convenient to use an oil quench. Naphthalene may be used as the quench medium,

[1] Badische Anilin und Soda Fabrik.

maintained at about 200 °C, and the heat input used to generate steam by means of a waste heat boiler. The cracked gases contain 8 per cent volume of acetylene. In the original Sachsse process the selective solvent for acetylene consisted of water under pressure, but more recent versions have switched to N-methylpyrrolidone. A feature of this process is that the by-product gas contains a substantial proportion of hydrogen and carbon monoxide. These gases, together with nitrogen from the air separation unit, form a suitable basis for ammonia production. The main difficulty lies in the inflexibility in supply of by-product raw materials. It is not always possible to strike an appropriate balance between acetylene and ammonia requirements. Nevertheless, the prospect of project integration of this kind is likely to prove a favourable factor in some cases.

A more recent development in this field is the S.B.A.[1] process. This employs the same principle as the Sachsse process, but uses a completely redesigned burner for the partial oxidation, and a modified quench assembly. The reaction temperature employed is said to be 1 400 °C. Acetylene is present in the cracked gases to the same extent as in the Sachsse process, about 8 per cent. Whilst the reaction of methane has been taken as the normal basis for these operations in the above outline some of the processes are flexible and may use propane or butane as satisfactory alternatives. Some reference has been made in the process description to the use of liquid hydrocarbon feedstocks. Further comment on this significant development belongs in the next section.

Liquid Hydrocarbons

Acetylene from Liquid Hydrocarbons

It has already been pointed out that both the Wulff and the Sachsse processes for acetylene manufacture have been adapted to the use of a naphtha feedstock. In practice both these processes have been found difficult to commission using a liquid feedstock, and the resolution of these problems may affect the process economics.

A series of acetylene processes has been developed specifically for liquid hydrocarbon feedstocks and with the additional feature that they are capable of providing ethylene as a co-product. These may be described as two-stage flame processes. Heat is provided by a burner

[1] Société Belge de l'Azote.

system using a fuel distinct from the feedstock, and commonly the off-gas separated from the process operations. The Hoechst H.T.P. process supplies the vaporized feedstock just beyond the top of a burner flame, raising the feed temperature to 1 200 °C for about 0.001 second. A water quench is applied immediately. The cracked gases may typically contain 10.6 per cent acetylene and 14.9 per cent ethylene.

A second process is a modification of the S.B.A. process using methane. Here again a vaporized naphtha is cracked in a reaction chamber where the heating medium is the product of combustion of the process off-gas with oxygen.

In both these processes high temperature steam is injected adjacent to the naphtha. A point of advantage for the liquid hydrocarbon feedstocks, in comparison with methane in acetylene production, is that an appreciably lower cracking temperature is feasible. Using naphtha it may be 1 100–1 200 °C in contrast to the 1 500 °C normally required for methane.

There are Japanese processes aimed specifically at cracking naphtha to both ethylene and acetylene for direct synthesis of vinyl chloride by the 'balanced process' (see p. 103).

A passing reference should be made to the proposal by B.A.S.F. to crack crude oil directly by a submerged flame technique. Oxygen is fed to the burner below the liquid surface. The flame creates an effective reaction temperature of about 1 500 °C and the surrounding oil at 200 °C acts as an automatic quench.

A commercial plant also exists in Japan for cracking 100 000 tons per year of crude oil to provide ethylene and acetylene—the combined yield of these being 46 per cent by weight of the feedstock. The reactor temperature is 900–1 200 °C. Here again the ethylene and acetylene are used for vinyl chloride production.

There is no shortage of proposals for variations in the technology of acetylene production. Many recent process developments are based on the initial treatment of hydrogen at very high temperatures. These 'plasma generators' are then used to direct the heat-treated hydrogen at a stream of appropriate hydrocarbon. A recent Hoechst announcement suggested that cracked gases at 1 000–1 200 °C from such a process should rapidly be quenched to 300 °C by high boiling hydrocarbons. Ethylene is likely to be a co-product of this type of operation. So far such proposals are mainly of technical interest.

An incidental point is that the present trend in European ethylene production is to use a high severity of naphtha cracking. This gives rise to an increasing acetylene content in the C_2 fraction. This acetylene normally has nuisance value in that it requires to be selectively hydrogenated to eliminate it from the stream. A recent suggestion has been that it may be worth recovering this acetylene by solvent extraction (e.g. by dimethylformamide). There could be about 5 000 tons per annum acetylene recoverable in this way from a 300 000 tons per year ethylene plant using naphtha feedstock, so that this is only likely to meet special situations requiring modest acetylene usage.

Overall acetylene is proving uneconomic as a large scale chemical feedstock. Its recovery, which seems rather unlikely, depends upon an economic breakthrough as a result of some major technological advance.

Naphtha Cracking for the Production of Lower Olefins

The essence of this operation is to start with a liquid hydrocarbon in the range naphtha to gas oil (but preferably a light naphtha), and to obtain from it, by cracking in the presence of steam, the maximum yield of useful products. These have been gradually extended from the original ethylene, propylene and butylenes, to include first butadiene and more recently benzene.

The liquid hydrocarbon feedstock will be a mixture of many compounds, but an empirical formula of the type CH_x may be postulated. As products ascend the scale of molecular weight the value of x is reduced. For a light naphtha the value of x may be about 2.2.

The cracking operation produces gases containing a higher proportion of hydrogen than the feedstock, and a liquid fraction with less hydrogen content. If the cracking is made too severe it is therefore apparent that trouble is likely to be experienced in the furnace tubes or the quench devices with coking.

The introduction of steam helps to minimize this coke formation by virtue of the water-gas reaction (see p. 181). The presence of steam also reduces the partial pressure of the oil vapour in the cracking zone. This is obviously desirable to achieve the maximum yield of gaseous products. There is also some indication that a secondary effect is to increase the proportion of olefins in the relevant gas streams.

It has already been pointed out that this operation differs from the thermal cracking carried out in the petroleum industry, in that it uses high temperatures and reduced pressures to effect a maximum yield of gases.

The feedstock is vaporized and rapidly heated, along with the steam, to the cracking temperature. Recent trends have been to use an increasing cracking temperature with the aim of raising the ethylene yield. Temperatures were once about 750 °C and current practice is moving towards a figure of 900 °C. The main effect of this is to raise the ethylene yield from a traditional 15–20 per cent by weight of the feedstock to 25–30 per cent.

Such figures must always be approximate in view of the wide variations in the feedstock itself. Of the hydrocarbons present in naphtha, the paraffins crack most readily to the lower olefins; the isoparaffins also crack readily but produce a higher proportion of methane; the naphthenes do not crack readily but can have the advantage of providing a relatively high yield of butadiene; the aromatics are scarcely cracked, and tend to emerge unchanged in the liquid fraction.

The effect of an increased cracking temperature, apart from increasing the ethylene yield, reduces the proportion of C_4 and liquid products, though it tends to increase the proportion of butadiene in the C_4 fraction, and slightly reduces the propylene yield. The cracking conditions must clearly be adjusted to the range of products required and the feedstock used. A further process variant is the possibility of recycling ethane from the distillation section to be cracked for additional ethylene production.

While any yield pattern is therefore a generalization from a picture of wide variation, typical indications of low and high severity cracking are given by the following figures. In each case the figure represents percentage of feedstock by weight, and it is assumed that ethane is recycled for cracking:

	Low Severity Cracking	*High Severity Cracking*
Ethylene	19.2%	32.3%
Propylene	14.6%	13.0%
C_4 stream	11.4%	8.5%
Fuel gas (propane and lighter)	12.8%	18.3%
Gasoline	36.3%	20.2%
Fuel oil	3.7%	5.7%
Loss	2.0%	2.0%

An average ethylene yield on feedstock today in Europe would be about 27 per cent and the figure is slowly rising. This reflects the trend towards a higher severity of cracking.

The naphtha is usually cracked in the presence of about 0.5 lb steam per lb hydrocarbon. It remains in the cracking zone for less than one second. The recent trend has been towards higher cracking temperatures applied for a shorter period of time.

The cracked products are rapidly quenched to prevent the cracking proceeding too far. A feature of modern designs is the improvement in devices to use the heat of the cracked products to generate maximum energy in the form of steam.

The feedstock normally passes through tube coils inside an oil-fired or gas-fired furnace. There is much refinement in the design of the furnace and coils in order to distribute the heat most effectively. With the more stringent temperature requirements there has been a switch from horizontal to vertical coils, together with some modification in materials of construction.

Many alternative methods of applying heat to the feedstock have been developed. These include heated pebbles, sand or refractory checker work, high-temperature steam, and partial oxidation techniques. In the tough competition of the present day, these methods appear no longer worthwhile, and new manufacturing projects in this field follow an increasingly uniform pattern.

Heavier feedstocks than naphtha have been used for olefin production. Gas oil cracking has been carried out where the economics are favourable—notably in the U.S.A. where naphtha has not yet become widely available for cracking. The cracking of gas oil may be carried out in equipment essentially similar to that used in naphtha cracking, but the olefin yields are rather lower than when cracking naphtha with comparable severity. In particular the cracking of gas oil offers less ethylene in relation to the co-products, so that the economic comparisons are quite complex.

When going a step further, and considering the cracking of crude oil directly to olefins, the inevitable problems of coking in tubular crackers make it necessary to carry out the heating in some form of fluidized bed. The heat input may arise from a partial combustion of the feedstock or by external heating of the solid particles introduced as a fluidized bed. The cracking temperatures are in the range of 720–850 °C and yields of over 20 per cent ethylene on the feedstock have been claimed. Processes along these lines have been developed particularly in Germany and Japan. They are unlikely to become of major significance unless some permanent shortage of naphtha can be foreseen.

Separation Processes

Cracking operations, whether based on refinery propane and ethane, or upon heavier liquid hydrocarbon feedstocks, produce a gas stream containing a mixture of constituents from hydrogen to the C_4 hydrocarbons, together with some C_5 and higher components in the form of a liquid fraction.

The first step in the separation of the cracked products is a primary fractionation which separates the liquid fraction (C_5 and higher). The separation of the C_4 and lighter components is a complex operation which normally involves refrigeration techniques and an elevated pressure so that most of the components can be treated in the liquid phase. Much ingenuity has been expended on the task of minimizing the capital and energy requirements of such an operation.

One technique, more common in Continental countries, is to take the gas stream to a modest pressure of about 12 atm., when the C_3 and C_4 components can be separated as liquids. The C_2 and lighter fraction is then taken down to a temperature of —140 °C in a 'cold box' (an insulated complex of equipment operating at low temperature) to separate the more volatile components from the C_2 compounds. This may involve a series of refrigerants such as liquid methane, ethylene and ammonia.

It is more usual to take the gases to a pressure of nearly 40 atm. The first separation step is the demethanizer column where methane and hydrogen are taken overhead, using a temperature of about —95 °C at the top of the column. This is maintained by using liquid ethylene to control the reflux temperature. Any ethylene carried overhead at this step is lost to the fuel gas stream, and any methane in the bottom product will remain as an impurity in the ethylene product. This is therefore a vital control factor in the whole operation. The bottom product from the demethanizer passes to the deethanizer column in which C_3 and C_4 hydrocarbons are the bottom product and C_2 components are taken overhead. At this point the

C_2 components are given a selective hydrogenation to remove acetylene before they pass to the final ethylene tower where ethylene is separated as an overhead product from ethane. The heavier components are further separated in a depropanizer column where the C_3 components are taken overhead, and a debutanizer column where the C_4 products are separated from some residual heavier material.

As the columns deal with larger molecules the pressure may be allowed to fall and the temperature to rise. The later columns may use liquid propylene or even water as the refrigerant.

It must be added that appropriate steps must be taken to keep these gas streams free of moisture, sulphur compounds, carbon dioxide and condensate.

The sequence of these separations may be varied, but the outline quoted is a typical arrangement.

There is a general trend towards more stringent specifications for the olefin products. This may call for additional acetylene removal steps and more precise fractionation. It is a normal requirement of modern plants that both the ethylene and propylene product should be 'polymer grade', that is, of purity 99.9 per cent. The separation of the C_4 components is discussed in Chapter 9.

A sensitive feature of this separation is the performance of the major compressors. Earlier designs used a multiplicity of reciprocating compressors. The current trend is to use single train centrifugal compressors of very large capacity. This has been of some importance in simplifying the design and minimizing the relative cost of the very large ethylene plants which are now conventional.

Wax Cracking

Wax cracking is a somewhat more specialized operation than the other processes already considered. Wax can be present to an embarrassing extent in some crude oils. This normally necessitates *inter alia* a solvent extraction process from which it is possible to recover the wax in an unrefined form. Some of this crude wax may be refined to be sold commercially as paraffin wax, but a substantial proportion is diverted to the production of chemicals.

The cracking of wax takes place at a temperature rather lower than that of most other cracking operations, of the order of 500 °C. The cracking proceeds for a period of time measured in seconds, usually in the presence of steam. A slightly elevated pressure is commonly employed.

The products of cracking, apart from a relatively small proportion of gas, and a fraction of fuel oil, comprise olefins in the range C_5–C_{25}. These are rarely separated into individual compounds, but appropriate fractions have been used as a feedstock to the Oxo process and as an alkyl fraction for alkylation processes usually designed to make detergent alkylates. Such wax-cracked olefins have found application in a variety of detergent products apart from detergent alkylate. The intermediates may be primary or secondary alcohols and the final products alkyl sulphates, ethoxylates or ether sulphates. In the production of synthetic detergents in the 1940's–1950's secondary alkyl sulphates, made by the direct sulphation of olefins, were important. Recently alkenyl sulphonates, made by a very mild sulphonation of C_{15}–C_{18} olefins have become among the more interesting products in this field.

The ideal product for most of these syntheses is represented by the straight chain alpha olefin. Even with purification treatments (e.g. by urea adducts) to minimize branch chains, the wax-cracked olefin products are by no means completely straight-chain, nor do they have all their double bonds in the alpha position. Such products are obtainable by the application of Ziegler techniques (q.v.) to ethylene, but such synthesized olefins are expensive. The choice, therefore, for these olefins, as happens so often, rests between essentially pure products at a relatively high price, or the less pure wax-cracked olefins at a lower price.

The production of straight chain olefins for detergent manufacture and other purposes is also carried out by dehydrogenation processes (sometimes involving some cracking) using appropriate liquid n-paraffins as feedstock.

The Oxo process is described on page 205, the production of detergent alkylate on page 138, and the use of Ziegler catalysts to make olefins on page 115. An assessment of the merits of different synthetic detergent types is given in Chapter 17.

Part 2

Petroleum Chemical Products and their Applications

A. CLASSIFICATION OF PETROLEUM CHEMICALS ACCORDING TO SOURCE

Chapter 4

Acetylene Derivatives

As a chemical raw material, acetylene is fighting a vigorous rearguard action, but it is emphatically under attack and its future is none too bright.

Acetylene from carbide has been particularly badly hit by attack both from hydrocarbon-based acetylene and from the various alternative chemical raw materials. An important carbide industry still exists in U.S.A., eastern Europe and Japan particularly, but the scale of production is being rapidly eroded.

The production of acetylene from hydrocarbons was actively developed in the U.S.A. in the early 1960's. Nearly all of this production was based on natural gas. In Europe there was some development along similar lines, supplemented in the middle and late 1960's by a number of units designed to produce acetylene from a naphtha fraction. The naphtha-based acetylene plants caused many difficulties and some were shut down after a very short life.

The position in the U.S.A. is changing so rapidly that the figures quoted for 1970, culled from various sources, may be slightly garbled:

End Product	*Usage of acetylene in '000 long tons*	
	1965	*1970*
Vinyl chloride	157	84
Acrylics (acrylates, acrylonitrile)	89	46
Polychloroprene	81	101
Vinyl acetate	62	96
Chlorinated solvents	67 (bracketed with the next two rows)	41
Acetylenic chemicals		18
Acetylene black		5
	456	391

These figures exclude non-chemical uses of acetylene.

Up to 1960 less than 20 per cent of U.S. acetylene was hydrocarbon-based and this rose to 43 per cent of the total by 1965. So far as chemical usage is concerned almost half the acetylene production was hydrocarbon-based by 1965. By 1970 the proportion was only slightly higher at about 54 per cent.

By 1975, however, the position in the U.S.A. will have undergone further radical change. The chemical usage of acetylene is expected by then to have dropped to about 200 000 tons and scarcely any of this will be from carbide.

This is a pattern that may well be seen elsewhere—though the movements may be more slowly accomplished.

Looking at the individual applications quoted in the table, the one that was biggest in 1965, vinyl chloride, has become less important because the main processes involved in this manufacture are now the chlorination and oxychlorination of ethylene.

In 1965 there was a substantial residual U.S. production of acrylonitrile from acetylene in spite of the increasing dominance of the propylene ammoxidation route. Today there is no U.S. production of acrylonitrile from acetylene, but in place of this there has developed a considerable production of acrylates and acrylic acid using acetylene as raw material. Here, as with acrylonitrile, it would seem that propylene is likely to become the dominant raw material.

Polychloroprene is now the largest U.S. user of acetylene. Already some of this production has switched to a butadiene basis and it is anticipated that by 1974 all U.S. polychloroprene will be made from butadiene.

Vinyl acetate has remained an important acetylene outlet in spite of competition from ethylene-based processes. The liquid phase reactions based on ethylene were not very successful, but the pace is now being set by the vapour phase processes using an ethylene raw material, and most new capacity is likely to use variants of this type of process.

Chlorinated solvents such as trichlorethylene and perchlorethylene have a long history of production from carbide acetylene. Companies continuing to use an acetylene raw material are increasingly turning to the hydrocarbon-based product. There has also been an important switch towards the use of ethylene as raw material (with ethylene dichloride as intermediate) in the U.S.A. though this trend is as yet less evident in Europe.

The one moderately bright spot in acetylene's future is the growth of 'Reppe chemistry'. This is a convenient collective term for the various products made by high pressure reaction of acetylene, and commemorates the pioneer work of the German chemist, Reppe, in this field. Acrylic acid is one such product that has already been mentioned, and others, all made on a relatively modest scale, include butyrolactone, butene diol, butyne diol and vinyl pyrrolidone. These are showing satisfactory growth but operate on a scale that does little to affect the overall gloom of the acetylene picture.

It is clear from the highly unsaturated nature of the acetylene molecule, that there will be little difficulty in persuading it to react. In practice, as with many other reactive compounds, the reactions of acetylene commonly use catalysts not to promote reactivity, but to direct it along one specific route, and to avoid side reactions. The purpose of such catalysts, in short, is selectivity.

In describing acetylene derivatives as petroleum chemicals we encounter a situation repeated many times in this book. The same products may be derived alternatively from petroleum sources or some other raw material, but the chemistry and technology of any further processing is normally the same whatever the raw material source.

Vinyl Chloride

The reaction of acetylene with hydrogen chloride

$$HC{\equiv}CH + HCl \rightarrow CH_2{=}CHCl$$

is carried out in the vapour phase using charcoal impregnated with about 10 per cent mercuric chloride as catalyst. The reagents must be pure and dry, with a slight excess of hydrogen chloride. The pressure used is atmospheric or slightly higher. The range of reaction temperatures quoted is a little varied, from 100–210 °C. This is not wholly unexpected with a highly exothermic reaction of this type, where reaction conditions must also be adapted to the age and condition of the catalyst.

Most of the vinyl chloride produced is polymerized to polyvinyl chloride for the manufacture either of resilient products such as sheeting or flooring materials, or rigid products such as pipes, etc. Vinyl chloride may be polymerized in various ways. Probably the most common technique is suspension polymerization. In this case, vinyl chloride droplets suspended in water are broken into smaller droplets by mechanical agitation, and a stabilizer (in the form of a protective colloid such as starch or polyvinyl alcohol) is added to minimize coalescence of the smaller droplets. The polymerization initiator used is soluble in vinyl chloride and may typically be lauryl peroxide or isopropyl percarbonate. The temperature of the reaction is about 55 °C.

Alternatively emulsion polymerization may be employed. Water is then the continuous phase, vinyl chloride the disperse phase; an emulsifying agent is used to stabilize the emulsion formed when the system is agitated. The polymerization initiator used is water-soluble.

Bulk polymerization of vinyl chloride can be carried out under pressure, with a temperature carefully controlled at about 60 °C. Solution polymerization is also feasible using a technique similar to suspension polymerization.

Each of the polymerization processes has some benefits. Emulsion polymerization, being continuously operable, is low in labour cost, and the product is easy to plasticize and process. The product from suspension polymerization is less affected by traces of auxiliary chemicals which can affect performance, and the consumption of auxiliaries is lower in this type of process. Bulk or mass polymerization yields a pure clear product and uses no auxiliary chemicals.

An associated product is vinylidene chloride. This may be made by reacting chlorine with either ethylene dichloride or vinyl chloride to give 1,1,2-trichloroethane. Hydrogen chloride is removed from trichloroethane by reacting with lime or caustic soda in slight excess at 98–99 °C. This reaction produces vinylidene chloride.

$$\underset{\text{vinyl chloride}}{\begin{matrix} CH_2 \\ \| \\ CHCl \end{matrix}} + Cl_2 \longrightarrow \underset{\text{trichloroethane}}{\begin{matrix} CH_2Cl \\ | \\ CHCl_2 \end{matrix}} \xrightarrow{+NaOH} \underset{\text{vinylidene chloride}}{\begin{matrix} CH_2 \\ \| \\ CCl_2 \end{matrix}} + NaCl + H_2O$$

Vinylidene chloride is co-polymerized with vinyl chloride to form products with special properties (e.g. Saran). The addition of hydrogen chloride to vinylidene chloride does not give 1,1,2-trichloroethane but 1,1,1-trichloroethane or methyl chloroform.

Methyl chloroform is now a product of some importance in its own right. In the U.S.A. consumption in 1970 was just over 150 000 tons and the forecast for 1975 is about 250 000 tons. It is largely used in metal degreasing, where its non-toxic properties give it a competitive advantage over trichloroethylene. It also has solvent applications in adhesives, polishes and speciality products. It has been suggested that by 1975/6 its usage in the U.S.A. will equal that of trichloroethylene. The use of methyl chloroform is also growing rapidly in Europe but the scale of development is some years behind that of the U.S.A.

The hydrochlorination of vinylidene chloride takes place over a Friedel-Crafts catalyst at 30 °C and a pressure marginally above atmospheric. The alternative process starting from ethylidene chloride is briefly described in Chapter 7.

The production of vinyl chloride from acetylene is now under very severe attack from the ethylene route (see Chapter 7). Traditionally the advantage of using acetylene lay in the fact that no by-product hydrogen chloride was formed, and the net usage of chlorine to make vinyl chloride from acetylene was less than that when ethylene was the raw material. The appearance of the oxychlorination techniques using ethylene overcame this problem. The main application of acetylene in this field lies in the operation of the 'balanced' process in which both acetylene and ethylene are used. These developments are discussed on pp. 103–4.

The proportion of vinyl chloride made in the U.S.A. from acetylene was dropping off gradually in the early 1960's. From 1960 to 1964 the percentage made from acetylene dropped from 54 to 45 per cent. This trend was accentuated in the later 1960's and by 1970 only about 14 per cent of U.S. vinyl chloride was still acetylene-based. The trend in Europe was a little slower to gain momentum, but by 1970 the effect was much the same.

In the U.S.A. the production of vinyl chloride was about 1.43 million long tons in 1970 (compared with 1.1 million long tons in 1966). Consumption there is expected to grow at 8–9 per cent a year. The United Kingdom does not publish figures for the monomer, but 314 000 long tons of polyvinyl chloride were produced in 1970.

The scale of vinyl chloride production is clearly indicated by the fact that the annual world capacity had reached 7.4 million tons by 1971 and was expected to be 11 million tons by 1975.

Vinylidene chloride is produced on a much more modest scale—probably rather more than 1 per cent of vinyl chloride production.

Vinyl Acetate

The normal reaction between acetylene and acetic acid is carried out in the vapour phase, using a temperature of about 210 °C over a catalyst comprising zinc or cadmium acetate on charcoal. The pressure is slightly above atmospheric. By using an excess of acetic acid the formation of ethylidene diacetate by-product is reduced to a minimum. The exit gases are cooled to 0 °C, at which temperature surplus acetylene is separated and recycled. Vinyl acetate is purified by distillation.

$$\underset{\text{acetic acid}}{CH_3COOH} + HC{\equiv}CH \rightarrow \underset{\text{vinyl acetate}}{CH_3COOCH{=}CH_2}$$

Vinyl acetate will polymerize readily, either alone or with other vinyl compounds. Polyvinyl acetate finds important applications in the production of adhesives and emulsion paints. The consumption of vinyl acetate in the U.S.A. was 415 000 long tons in 1970 and is forecast to rise to about 650 000 tons in 1975. In the United Kingdom there was a capacity limitation holding production to about 25–30 000 tons compared with a demand rising to near 50 000 tons in 1971. A new plant of capacity 50 000 tons per year vinyl acetate (based on ethylene feedstock) has recently been built in the United Kingdom.

Once again ethylene competes with acetylene as the raw material. The original liquid phase process for making vinyl acetate from ethylene has not proved successful. New vinyl acetate capacity is largely based on the vapour phase reaction of ethylene and acetic acid. This is discussed in Chapter 7.

Polychloroprene Rubbers

These are synthetic rubbers which are notable for resistance to deterioration in the presence of aliphatic chemicals and hydrocarbons.

The first step in the traditional production of polychloroprene rubbers is the liquid phase dimerization of acetylene to monovinylacetylene. This is carried out in the presence of an aqueous solution of cuprous chloride and ammonium chloride at 65–75 °C and marginally above atmospheric pressure. With a contact time of 10–15 seconds the conversion is about 20 per cent per pass, and the yield 60–65 per cent.

$$2HC{\equiv}CH \rightarrow \underset{\text{monovinylacetylene}}{HC{\equiv}CCH{=}CH_2}$$

Monovinylacetylene will add a molecule of hydrogen chloride in the presence of aqueous cuprous chloride at 30–60 °C to give chloroprene. The yield is over 90 per cent on the monovinylacetylene.

$$\underset{\text{monovinylacetylene}}{HC{\equiv}CCH{=}CH_2} + HCl \rightarrow \underset{\text{chloroprene}}{CH_2{=}CClCH{=}CH_2}$$

Polychloroprene is produced by an emulsion polymerization process. In one type of operation the polymerization is radical-initiated (using, for example, a persulphate) with mercaptans used to control the molecular weight of the polymer. Alternatively a complex procedure involving copolymerization with sulphur may be used.

The production of polychloroprene rubbers is much more extensive in the U.S.A. than elsewhere. The 1970 U.S. production figure was about 165 000 long tons, and of this about 30–35 000 tons was exported. The consumption in the whole of western Europe is scarcely half the U.S. production. Capacity for polychloroprene rubbers was marginally above 100 000 tons in 1971 and the annual capacity should rise to perhaps 160 000 tons in western Europe by 1975.

Traditionally polychloroprene has been made from acetylene, but the competitive process based on a butadiene raw material is gaining ground. By 1974 it is expected that all U.S. polychloroprene will be butadiene-based, and a number of such plants will be in operation elsewhere.

United Kingdom production is represented by the Northern Ireland plant of du Pont, which is supplied with acetylene by an adjacent Wulff plant, producing its acetylene from a naphtha feed-stock. The U.K. production figure is not disclosed, but is of the order of 20–30 000 tons per year.

Trichloroethylene

This is an operation which until recently has mainly used acetylene from coal. Acetylene is chlorinated in tetrachloroethane solution at 80 °C and at a reduced pressure, using antimony or ferric chloride as a catalyst. Acetylene and chlorine are introduced independently into the liquid tetrachloroethane. The reaction is highly exothermic, and provision for cooling the reaction products must be made. A gas phase reaction is possible but is not operated in practice.

The chlorinated product, 1,1,2,2-tetrachloroethane, is dehydrochlorinated to give trichloroethylene. The removal of hydrogen chloride may be achieved by reacting tetrachloroethane with a slurry of lime at the boil, but this occasions some loss of chlorine as calcium chloride. There is some preference today for a thermal dehydrochlorination which may use a catalyst (such as barium chloride on charcoal) at temperatures of 250 °C upwards, or may operate without a catalyst, in which case the temperature used is likely to be 500 °C or more.

$$HC{\equiv}CH + 2Cl_2 \longrightarrow \underset{\text{tetrachloroethane}}{CHCl_2{-}CHCl_2} \xrightarrow{-HCl} \underset{\text{trichloroethylene}}{CCl_2{=}CHCl}$$

The production of perchloroethylene from trichloroethylene is a two stage process following closely the pattern of trichloroethylene production itself. Trichloroethylene is first chlorinated in pentachloroethane solution at 80–90 °C, using a metal chloride catalyst, to give pentachloroethane. This is dehydrochlorinated to perchloroethylene using reaction with alkali or pyrolysis techniques to remove the hydrogen chloride.

$$\underset{\text{trichloroethylene}}{CCl_2{=}CHCl} \xrightarrow{+Cl_2} \underset{\text{pentachloroethane}}{CCl_3{-}CHCl_2} \xrightarrow{-HCl} \underset{\text{perchloroethylene}}{CCl_2{=}CCl_2}$$

The production and consumption of trichloroethylene in the U.S.A. was almost in balance in 1970 at 273 000 long tons. In the countries of the European Economic Community the production in 1970 totalled 246 000 tons. Future growth, both in the U.S.A. and Europe, is likely to be limited to about 3 per cent per year. About 95 per cent of trichloroethylene usage is for metal degreasing (the figure is similar in U.S.A. and Europe) with a minor application as an extraction solvent. A possibility for a substantial new market for trichloroethylene (together with other chlorinated solvents) could arise if certain textile processes, at present carried out in an aqueous medium, became solvent-based. The original trichloroethylene application in dry cleaning has been taken over almost entirely by perchloroethylene.

Perchloroethylene was produced in 1970 to the extent of 313 000 long tons in the U.S.A. and 212 000 tons in the Common Market countries. There is a net export from both these areas. In the U.S.A. the perchloroethylene demand for 1970 was detailed as follows:

Dry cleaning solvent	73%
Chemical intermediate	8%
Vapour degreasing	7%
Export and other	12%

The use of perchloroethylene as a chemical intermediate is largely in the production of fluorocarbons. The hydrofluorination of perchloroethylene, for example, provides 1,1,2-trichloro,1,2,2-trifluoroethane (which oddly abbreviates to chlorofluorocarbon 113), a versatile solvent.

In the past few years there has been rapid development of alternative processes for trichloroethylene and perchloroethylene made by the chlorination or oxychlorination of ethylene. In addition perchloroethylene (in conjunction with carbon tetrachloride) may be produced by chlorinolysis of various hydrocarbons (see Chapter 6).

Acrylonitrile

Acrylonitrile has had a short history and in the late 1950's it represented a major development of acetylene chemistry. The reaction involved is the addition of acetylene to hydrogen cyanide.

$$HC{\equiv}CH + HCN \longrightarrow \underset{\text{acrylonitrile}}{CH_2{=}CHCN}$$

The process operates at atmospheric pressure and a temperature of around 70–90 °C, in the presence of a catalyst comprising a solution of cuprous chloride in hydrochloric acid. The propylene ammoxidation processes have now become wholly dominant as sources of acrylonitrile. The developing importance of acrylonitrile is therefore further considered following a description of these processes in Chapter 8.

Acetaldehyde and its Derivatives

One of the ways of producing acetaldehyde is by the hydration of acetylene, using sulphuric acid in the presence of a catalyst comprising a reduction–oxidation system of mercurous/mercuric sulphate buffered by ferric sulphate.

$$HC{\equiv}CH + HOH \rightarrow \underset{\text{vinyl alcohol}}{H_2C{=}CHOH} \rightarrow \underset{\text{acetaldehyde}}{CH_3CHO}$$

The initial product of the reaction is vinyl alcohol, which promptly re-arranges itself to give acetaldehyde. This reaction takes place at roughly atmospheric pressure and 95 °C in a tower. Only about 53 per cent of the acetylene fed is reacted and the remainder is recycled, after removal of organic products. This route to acetaldehyde has largely been replaced by the ethylene routes, either using direct oxidation or via ethyl alcohol. The acetylene basis is still used to a slight extent in Europe, but has been abandoned in the U.S.A.

Acrylic Acid and Acrylates

There are many routes to acrylic acid, but one of the more notable employs the reaction between acetylene, carbon monoxide and water. This is a high pressure reaction using a combination of nickel and halogen (e.g. nickel bromide) as catalyst. Typically the operation may proceed in tetrahydrofuran solution at a pressure of 60–200 atm.

$$C_2H_2 + CO + H_2O \rightarrow CH_2{=}CHCOOH$$

A similar process may operate using an alcohol in place of water, in which case an acrylate ester is formed directly.

Other routes to acrylic acid and the acrylates operate by way of ethylene cyanohydrin (see Chapter 7), propylene oxidation (see Chapter 8), or the reaction between ketene (see Chapter 7) and formaldehyde to give 2-propiolactone which gives an acrylate on addition to an appropriate alcohol.

Isoprene

There are many prospective isoprene processes. One which has just achieved commercial operation is the Italian process based on the reaction of acetylene and acetone. This reaction takes place at 20 atm. and 10–40 °C using liquid ammonia as solvent, in the presence of an alkaline catalyst. The methyl butynol formed in this reaction (as a water azeotrope) is hydrogenated to methyl butenol. This dehydrates over alumina at 260–300 °C to isoprene.

$$C_2H_2 + (CH_3)_2CO \rightarrow CH_3C(CH_3)(OH)C{\equiv}CH$$

acetylene, acetone → methyl butynol

$$\downarrow +H_2$$

$$CH_2{=}C(CH_3)CH{=}CH_2 \xleftarrow{-H_2O} CH_3C(CH_3)(OH)CH{=}CH_2$$

isoprene ← methyl butenol

A 32 000 tons per annum plant using this process has been built in Italy. The economics of this route to isoprene remain unproven, but there are some outlets developing for the intermediate products.

Acetylenic Chemicals

A major chemical development attributable to Reppe in Germany arose from the technology of reacting acetylene under pressure with aldehydes, ketones and alcohols as well as carbon monoxide (see Acrylic Acid above). A range of products of some industrial importance stems from the reaction of acetylene with formaldehyde to form 1,4-butyne diol ($CH_2OHC{\equiv}CCH_2OH$). This may be successively hydrogenated to the corresponding butene diol and butane diol. The latter may be dehydrogenated to give butyrolactone. This will react with ammonia under pressure to give pyrrolidone. The chain of reactions may be taken one stage further by reacting pyrrolidone with acetylene under pressure to give vinyl pyrrolidone.

These products enter the acetylene usage statistics under the heading 'acetylenic chemicals'.

Chapter 5

Methane Derivatives

Acetylene See page 27.
Ammonia See page 180.
Methyl Alcohol See page 197.
Carbon Black See page 252.

The above derivatives are discussed elsewhere as indicated. Acetylene is treated for convenience as a raw material, and its derivation is, therefore, discussed in the chapter on raw materials. Ammonia, methyl alcohol and carbon black can be derived not only from methane, but from a wide range of alternative hydrocarbons. They are, therefore, excluded from the classification according to source, and examined separately in the subsequent product groups.

Hydrogen Cyanide

This is a long-established product that developed very rapidly in the late 1950's and early 1960's, with acrylonitrile as a major pacesetter. The scope for development has been significantly reduced with the switch of acrylonitrile to a propylene feedstock basis.

The main petroleum route to hydrogen cyanide is the reaction between hydrocarbons (in practice essentially methane) and ammonia —the Andrussow process. This process requires the use of oxygen in addition to methane and ammonia. In earlier days methane from coke oven gas was treated in the same way.

$$2NH_3 + 3O_2 + 2CH_4 \leftrightharpoons 2HCN + 6H_2O$$

Ammonia, air and natural gas (with higher paraffins usually removed) are passed over a platinum-rhodium catalyst at a pressure slightly above atmospheric and a temperature of about 1 000 °C. A consideration of reaction equilibrium would suggest a higher temperature still, but at such high temperatures product decomposition is too rapid. The reacted gases are cooled, and scrubbed with acidified water to remove ammonia. The hydrogen cyanide is then absorbed in water, and purified by distillation.

Today, significant quantities of hydrogen cyanide are recovered as a by-product from acrylonitrile plants using the propylene ammoxidation process.

The major earlier process involved the production of sodium cyanide from sodamide and carbon, and the subsequent acidification of sodium cyanide.

$$\underset{\text{sodamide}}{NaNH_2 + C} \xrightarrow{-H_2} \underset{\text{sodium cyanide}}{NaCN} \xrightarrow{H_2SO_4} HCN$$

Now the reverse process is used, making sodium cyanide from hydrogen cyanide.

The demand for hydrogen cyanide in the U.S.A. has fluctuated as follows:

1947	4 500 long tons
1959	84 500 long tons
1964	168 000 long tons
1970	121 000 long tons
1971	129 000 long tons
1975 (est.)	152 000 long tons

The later figures are not completely comparable with the earlier ones, since they apparently omit reference to the du Pont usage of hydrogen cyanide for adiponitrile manufacture. A recent breakdown of usage (again omitting du Pont adiponitrile) gave these percentages in the U.S.A.:

Methyl methacrylate	61%
Sodium nitrilotriacetate and other chelating products	21%
Sodium cyanide	10%
Others	8%

Up to the middle 1960's acrylonitrile was the major outlet for hydrogen cyanide in the U.S.A.

Recently great hopes were held of a vast market for hydrogen cyanide in the production of sodium nitrilotriacetate. This was proposed as a detergent 'builder' to replace the conventional polyphosphates, which were claimed in certain circumstances to create pollution problems.

There are a variety of processes for making nitrilotriacetic acid, and the more favoured ones involve hydrogen cyanide, formaldehyde and ammonia as raw materials. The requirements, on an almost theoretical basis, are 0.5 ton formaldehyde, 0.45 ton hydrogen cyanide and 0.1 ton of ammonia per ton of acid.

The conventional process involves the reaction of all three raw materials to form nitrilotriacetonitrile.

$$NH_3 + 3CHOH + 3HCN \longrightarrow N(CH_2CN)_3 + 3H_2O$$

The nitrile is then hydrolysed to the acid $N(CH_2COOH)_3$. Alternatively by varying the raw material steps, the intermediate may be formaldehyde cyanohydrin or hexamethylene tetramine. The form of product used as a detergent builder is the trisodium salt of nitrilotriacetic acid.

Prospects for such a development were dashed, for the time being at least, by the indication that sodium nitrilotriacetate represented a toxicity hazard.

Processes for hydrogen cyanide production, similar to the Andrussow process, but omitting the use of oxygen, are also operated commercially.

Degussa employs a platinum catalyst in the form of a thin coating on the inside of ceramic tubes. The tubes are externally heated and the reaction takes place at 1 200–1 300 °C. The hydrogen produced is relatively pure.

The Shawinigan process operates without a catalyst but in the presence of a fluidized bed of heated coke at 1 500 °C. Here the hydrocarbon feedstock may range from methane to a naphtha fraction.

$$CH_4 + NH_3 \longrightarrow HCN + 3H_2$$

A process which was once of significance was the reaction of carbon monoxide with methyl alcohol to give methyl formate. This will react with ammonia to give formamide and release methyl alcohol. Formamide dehydrates to hydrogen cyanide.

Chlorinated Methanes

All the chlorinated methanes may be obtained by processes not directly involving methane. Methyl chloride is obtained by the reaction of methyl alcohol and hydrogen chloride, and may then be

Petroleum chemical production

Courtesy Shell Chemical Co. Ltd

Fine chemical production

Courtesy Bush Boake Allen Ltd

Both in technique and equipment the requirements of petroleum chemical manufacture are different from those appropriate to the production of fine chemicals

Ethylene plant in Czechoslovakia—the low temperature gas separation

Courtesy Humphreys & Glasgow L

chlorinated to methylene dichloride. Chloroform has been made by the reduction of carbon tetrachloride and the reaction of bleaching powder with ethyl alcohol or acetone. For carbon tetrachloride, two alternative processes are of some importance. The first is the chlorination of carbon disulphide, and the second is the co-production of carbon tetrachloride and perchloroethylene by the thermal chlorination of hydrocarbons (see Chapter 6). Some carbon tetrachloride also occurs as a by-product of trichloroethylene production.

All four chlorinated methanes are now produced to an increasing extent by the direct chlorination of methane.

Methane is the most difficult of the paraffins to chlorinate. Since methyl chloride itself chlorinates faster than methane, a large excess of methane must be used if a reasonable yield of methyl chloride is to be obtained. As an illustration of this, a reaction between 5 mols of methane and 1 mol of chlorine at 400 °C will produce a reaction mixture of 50 per cent methyl chloride, 35 per cent methylene dichloride and 15 per cent of more highly chlorinated compounds.

$$CH_4 + Cl_2 \rightarrow CH_3Cl + HCl$$
$$CH_3Cl + Cl_2 \rightarrow CH_2Cl_2 + HCl$$

If the reaction is required to proceed essentially only as far as methyl chloride a methane/chlorine mol ratio of 10 : 1 is needed. Where this mol ratio is reduced to 1.7 : 1 a reasonable balance of all four chlorinated methanes is obtained.

Obviously the proportion of the more highly chlorinated products can be increased by recycling the less chlorinated methanes for further chlorination. Excessive levels of recycle may, however, lead to carbon formation on an unacceptable scale.

The chlorination may proceed at temperatures within the range 350–450 °C according to individual circumstances, and at pressures marginally above atmospheric. Thermal chlorination is common. A catalyst which has been used is partially reduced cupric chloride deposited on pumice. In some instances it is preferred to promote the reaction by irradiation with ultra-violet light.

The production of methyl chloride, by the reaction of methyl alcohol vapour and hydrogen chloride, takes place at 350 °C, in the presence of a catalyst such as calcined alumina or zinc chloride on pumice.

$$CH_3OH + HCl \rightarrow CH_3Cl + H_2O$$

Methyl chloride production in the U.S.A. has risen rapidly from slightly over 50 000 long tons in 1961 to almost 200 000 tons in 1970. A more modest growth, to about 255 000 tons in 1974, is forecast. Some 38 per cent is used as a chemical reagent in the production of silicone resins, 38 per cent is for tetramethyl lead, 5 per cent for butyl rubber, 4 per cent for methyl cellulose and 15 per cent finds other applications, including the production of quaternary ammonium compounds. The pace of methyl chloride growth has been set by the expansion of tetramethyl lead as a gasoline additive. It seems likely that gasoline legislation in the U.S.A. will affect this application in the future.

Methylene dichloride has achieved an expansion in the U.S.A. comparable to that of methyl chloride. Consumption, which was about 60 000 tons in 1960, reached over 190 000 long tons in 1970. A growth rate of 8 per cent a year is forecast for the next few years. The applications for U.S. methylene dichloride have recently been detailed as follows:

Paint remover	40%
Export	20%
Solvent degreasing	10%
Aerosol vapour pressure depressant	8%
Plastics processing	6%
Miscellaneous	16%

Some of the export demand is for war-related paint removal. This factor, added to the trend towards expanding capacity in countries other than the U.S.A., is likely to cause a reduction in methylene dichloride exports from the U.S.A. The plastics processing application is largely as a heat control agent in the acetylation of cellulose to produce cellulose acetate.

Chloroform is the least substantial of the chlorinated methanes in terms of production, but in recent years growth has been spurred by the development of fluorocarbons. The U.S. demand for chloroform in 1970 was about 107 000 long tons, and a growth of about 10 per cent a year is forecast. Over 50 per cent of chloroform is used for fluorocarbon refrigerants and propellants, and a further 41 per cent is for fluorocarbon plastics.

Carbon tetrachloride is the biggest of all the chlorinated methanes in scale of production and use. As with chloroform, the dominant application is for fluorocarbons. The production of dichlorodifluoromethane accounts for 69 per cent of the demand, and trichlorofluoromethane for another 26 per cent. Earlier applications, such as dry cleaning, degreasing and grain fumigating are now of negligible importance. The demand for carbon tetrachloride in the U.S.A. was about 418 000 long tons in 1970, and a growth of perhaps 6–7 per cent a year is envisaged.

There is a variant on the two-stage operation whereby carbon tetrachloride is first produced and purified so that it may be fluorinated to the appropriate fluorocarbons. Montedison operates a process in Italy on a modest (13 000 long tons per year) scale, which produces a 50–50 mixture of dichlorodifluoromethane and trichlorofluoromethane by the direct halogenation of methane.

$$CH_4 + 4Cl_2 + HF \rightarrow CCl_3F + 5HCl$$
$$CH_4 + 4Cl_2 + 2HF \rightarrow CCl_2F_2 + 6HCl$$

The reaction takes place at 370–470 °C and 4–7 atmospheres. The fluidized catalyst is understood to be a mixture of metallic chlorides or fluorides on an appropriate support. The hydrogen chloride passes off for use in other processes. The product mixture is given a water wash and a caustic wash followed by a sulphuric acid dehydration. The individual fluorocarbons are purified and separated in a two-column distillation unit.

An important aspect of the process is the control of reaction temperature, and improvement in yield and selectivity, by adjustment of a recycle stream which includes carbon tetrachloride.

Carbon Disulphide

Carbon disulphide was once largely produced by the direct reaction of charcoal and sulphur. The process now preferred is based on methane and sulphur. Vaporized sulphur is reacted with methane of high (99 per cent) purity. The reactor is a high-chrome steel vessel filled with a catalyst of activated alumina or synthetic clay. The conditions are somewhat drastic, 670–680 °C and a slightly elevated pressure of about 2 atm. This enables the conversion to reach about 85 per cent on methane and over 90 per cent on sulphur. Earlier processes used less drastic conditions, but in view of the lower conversion, it was found necessary to operate a methane recycle.

The main reactions are:

$$CH_4 + 2S_2 \rightarrow CS_2 + 2H_2S$$
$$CH_4 + S_2 \rightarrow CS_2 + 2H_2$$
$$CH_4 + 2H_2S \rightarrow CS_2 + 4H_2$$

Unreacted sulphur is removed from the reaction gases by scrubbing with liquid sulphur, and crude carbon disulphide is separated by absorption in a hydrocarbon solvent. The off-gas from the absorber, which may be 90 per cent hydrogen sulphide, is normally treated for sulphur recovery.

As methane is by no means universally available, alternative processes reacting sulphur with fuel oil or petroleum coke have been proposed, but they have not yet been developed to a significant extent.

The demand for carbon disulphide in the U.S.A. was approximately 366 000 long tons in 1970. This product is showing a relatively slow growth (in petroleum chemical terms) of about 3 per cent a year. This is not surprising since the main outlet, taking 65 per cent of the total, is in viscose rayon and cellophane manufacture. A further 28 per cent is used for the production of carbon tetrachloride, where the direct chlorination of methane and a variety of alternative chlorination processes offer competition. The remaining 7 per cent of carbon disulphide finds miscellaneous uses.

Chapter 6

Derivatives of Higher Paraffins

The main raw materials considered in this section are ethane, propane and butane. The major applications of ethane and propane for pyrolysis to ethylene have already been discussed. Ethane, propane and butane can be oxidized to synthesis gas (for ammonia or methyl alcohol production) in common with a wide variety of other hydrocarbons.

In this section it is proposed to deal essentially with those products and processes with a specific application to the paraffins quoted. Nitroparaffins are included here for convenience, although the methane derivative is one of the series. Some reactions of paraffins in the C_{10}–C_{18} range are included as a part of the background to the story of synthetic detergent development.

Ethyl Chloride

Most ethyl chloride is made by hydrochlorination of ethylene, but there are two processes in commercial use involving the chlorination of ethane. This is somewhat easier to perform than the chlorination of methane.

In practice, ethane is chlorinated, using a large excess of ethane, at a temperature of 300–500 °C. Since ethyl chloride chlorinates much more slowly than ethane, it is relatively easy to direct the reaction towards the monochlorinated product.

$$C_2H_6 + Cl_2 \rightarrow C_2H_5Cl + HCl$$

A practical disadvantage of this process may be the formation of by-product hydrogen chloride, for which there is not always a ready demand. This operation has therefore been modified, to combine the reaction above with the hydrochlorination of ethylene. It was found that in the gas phase the chlorination of ethane would proceed at 400 °C in the presence of ethylene without any serious development of addition reactions. By this means, the chlorine is completely utilized, with a minimum of by-product formation.

$$\underset{\text{ethane}}{C_2H_6} + Cl_2 \rightarrow \underset{\text{ethyl chloride}}{C_2H_5Cl} + HCl$$

$$\underset{\text{ethylene}}{C_2H_4} + HCl \rightarrow \underset{\text{ethyl chloride}}{C_2H_5Cl}$$

This process is the one used for the production of ethyl chloride in the United Kingdom.

Of the ethyl chloride consumption in the U.S.A., which amounted to nearly 270 000 long tons in 1970, 90 per cent was consumed in the manufacture of tetraethyl lead, which is added to gasoline to improve the anti-knock rating. The remaining 10 per cent is for various uses, including the manufacture of ethyl cellulose, and application as refrigerant and anaesthetic.

This product is relatively unusual in that its usage has begun to decline. This follows from its almost complete dependence on the lead additive market for gasoline which is a target for attack on the grounds of pollution. The main objection is not pollution of the air by lead itself. The pollution which arises from gasoline comprises carbon monoxide, unburned hydrocarbons and oxides of nitrogen in car exhausts. The intention in the U.S.A. is to control these emissions by the use of a catalytic 'after-burner' in the car exhaust system and the presence of lead in gasoline seems likely to interfere with the catalytic activity of the after-burners. It may be mentioned that on this subject all is hypothesis (and emotional hypothesis at that). The effectiveness of the catalytic 'after-burner' has yet to be established.

The market for ethyl chloride in the U.S.A. will probably continue to decline. Various forecasts range between 20 per cent and 90 per cent of the current figure as the market for 1975.

Vinyl Chloride from Ethane

This process was announced in 1971 on the strength of pilot plant work. It has yet to be commercialized. The ethane is said to be converted in a single step to vinyl chloride monomer, the single step embracing the operations of chlorination, oxychlorination and dehydrochlorination. It is further claimed that no by-products are involved. This somewhat spectacular achieuement is said to arise from the use of a unique molten-salt catalyst system.

The merit of such a process could be considerable where ethane is cheap and abundant, conditions likely to arise where 'wet' natural gas is plentiful. Even in the U.S.A. these circumstances are becoming scarcer, and in Europe such a process development is likely to be considerably hampered by limited availability of a suitable feedstock.

Nitroparaffins

These are products of potential, rather than of major, commercial importance at this stage.

The commercial development of this range rests largely upon the vapour phase nitration of propane with nitric acid. In the reactor excess propane is treated with sprays of 75 per cent nitric acid at 8 atmospheres pressure and 400–450 °C. The nitro-products obtained will normally be in the range 10–30 per cent nitromethane, 20–25 per cent nitroethane, 55–65 per cent nitropropanes. The separation is effected in a series of distillation columns. Amongst the lower paraffins, the smaller the molecule, the more difficult is the nitration.

It is believed that the reaction proceeds by a free-radical mechanism (the chlorination and oxidation reactions of paraffins are understood to be similar in this respect). The conversion to nitroparaffins may be significantly increased, therefore, by the presence of oxygen in the reactor as a source of free radicals.

Nitroparaffins have found a variety of applications in the propellant field, as chemical intermediates and as solvents. Their usage is increasing quite rapidly. As production is virtually limited to one company in the U.S.A., detailed figures of usage are rarely available. It has been estimated that the consumption of nitroparaffins in 1969 was about 20 000 tons in the U.S.A. and 1 000 tons in Europe.

Chlorination of Propane

The term 'chlorinolysis' has been applied to the intensive thermal chlorination of hydrocarbons. The hydrocarbons involved may be up to C_3 and are typified here by the reaction of chlorine on propane. In fact both paraffins and olefins may be treated in this way.

The products of such a reaction are carbon tetrachloride, perchloroethylene and hydrogen chloride. This is a vapour phase reaction which requires no catalyst. The process is said to be flexible in that the proportions of perchloroethylene and carbon tetrachloride produced can be varied widely, since additional perchloroethylene can be made by recycling carbon tetrachloride to the furnace.

The reaction, which typically proceeds at a temperature of 590 °C and a pressure very slightly above atmospheric, may be expressed in an over-simplified form as:

$$C_3H_8 + 8Cl_2 \rightarrow CCl_4 + CCl_2{=}CCl_2 + 8HCl$$

A very similar operation, with a totally different process background, is the thermal cracking of propylene dichloride. This compound occurs as a fairly useless by-product of propylene oxide production by the chlorohydrin route (Chapter 8). The economics of such a cracking operation are only favourable where a substantial quantity of propylene dichloride can be accumulated in one place. The main products of the cracking are carbon tetrachloride and perchloroethylene.

Ammoxidation of Propane

Details of this process, not yet in the commercial stage, are somewhat sparse. Propane, ammonia and air are passed over a catalyst. The process is similar in essence to the well-known ammoxidation of propylene (Chapter 8), the end-product being acrylonitrile and the by-products acetonitrile and hydrogen cyanide. It differs from the operation based on propylene in that no steam is used as a diluent.

It is said that propane of 92–98 per cent purity is preferred for this reaction. This limits the application of the process, since refinery propane is not available at this concentration without prior purification. Probably the main application would lie in areas where adequate supplies of 'wet' natural gas are available.

Oxidation of Propane and Butane

In the normal operation of an industrial chemical process one goes to immense pains to ensure that one or two reactions form the basis of the production, so that subsequent purification is kept as simple as possible. An exception to this rule is the vapour phase oxidation of propane and butane with air, as it is carried out in the U.S.A.

The oxidation takes place at a temperature of about 370 °C and pressures of 7–8 atmospheres. No catalyst is used.

A switch to oxygen as the oxidizing medium some years ago was said to increase the flexibility and capacity of the unit. In spite of this, some subsequent capacity used air oxidation.

Something like 15–20 per cent of the hydrocarbon is lost as oxides of carbon. The remaining product mixture includes formaldehyde, methyl alcohol, acetaldehyde, acetic acid, n-propyl alcohol, methyl ethyl ketone, acetone and other oxygenated chemicals.

To illustrate the importance of reaction conditions on propane oxidation, it may be noted that under 1 atm. pressure at 350 °C the major oxidation products are formaldehyde, acetaldehyde and methyl alcohol. When the temperature is raised above 400 °C, the production of propylene and hydrogen peroxide becomes important. At temperatures above 450 °C, the proportions of oxygenated products diminish further and the major products are propylene, ethylene, methane and hydrogen.

Similar effects are seen with vapour phase oxidation of butane. At relatively low temperatures the major products are acetaldehyde, formaldehyde and methyl alcohol. At about 375 °C the butylenes become a major product and on further increase of temperature propylene and ethylene become the dominant products.

In practice it is the oxygenated products that are called for and the reaction conditions are chosen accordingly. The problems of separating the pure products are formidable. This is a process which can scarcely be described as elegant, but which becomes commercially feasible by virtue of a simple reaction step and an economical raw material. This is perhaps an extreme case, for the recent trend (outlined below) is towards a greater degree of product control.

The liquid phase oxidation of butane, which is now assuming importance, is a more selective operation. If a catalyst is used it is likely to be cobalt or manganese acetate. Celanese Corporation use such a catalyst and operate at 54 atm. pressure and 175 °C. Hüls, when they operated a process of this type, did not appear to use a catalyst and their operating conditions were 170–200 °C and 65 atm. It is necessary to use a solvent to maintain a liquid butane phase. The main product of this reaction is acetic acid. Other oxygenated products are said to include formic acid, lower alcohols, acetone and methyl ethyl ketone.

If the oxidation conditions chosen involve a lower temperature and accept a lower conversion, the C_4 carbon skeleton of the feedstock may be retained, and the main oxidation product is then methyl ethyl ketone, together with some secondary butyl alcohol.

These operations account for about 16 per cent of the formaldehyde capacity, 5 per cent of the methyl alcohol capacity and over half the acetic acid capacity in the U.S.A.

Butadiene from Butane

Butadiene is one of the vital components of the giant synthetic rubber industry (discussed in more detail on page 172). Its production is today one of the largest scale organic chemical operations. Most butadiene in the U.S.A. is made by the dehydrogenation of butylenes or n-butane. In Europe, where ethylene is produced by naphtha cracking, butadiene is a valuable component of the C_4 gas stream produced.

The production of butadiene by the dehydrogenation of butane is an important process. Butylenes are formed as a by-product.

$$\underset{\text{butane}}{C_4H_{10}} \rightarrow \underset{\text{butadiene}}{CH_2{=}CHCH{=}CH_2} + 2H_2$$

$$\underset{\text{butane}}{C_4H_{10}} \rightarrow \underset{\text{butylene}}{C_4H_8} + H_2$$

The dehydrogenation takes place at 600–650 °C and 0.2 atmospheres in a reactor filled with chromia-alumina catalyst. (This illustrates the general principle that for a reaction involving an increase in product volume, the conversion will be increased by operation under reduced pressure, or with the partial pressure reduced by the introduction of a diluent.) The catalyst becomes covered in carbon after perhaps 8–10 minutes. The feed mixture is then directed to another reactor (which has been brought up to cracking temperature) and the carbon is burnt off the catalyst of the first reactor with preheated air. This regenerates the catalyst, after which the reactor is steam-purged to eliminate combustion gases. A minimum of three reactors in parallel is therefore required for continuous operation: one on stream, one being regenerated, and one being steam-purged. The reactor effluent is oil-quenched, stripped of its volatile fraction ('light ends') and the butadiene separated by one of the various methods outlined in Chapter 9. Normal fractional distillation cannot be applied to the mixture of C_4 hydrocarbons, since the boiling points are so close together.

In 1970 the proportion of U.S. butadiene produced from n-butane was about 30 per cent. The proportion is tending to go down as butadiene becomes increasingly available in the U.S.A. from the C_4 streams derived from liquid-feed ethylene plants.

It may be noted that facilities designed for butane dehydrogenation may be adapted to use a butylene feedstock where the economics are favourable. The advantage is that a given facility is immediately granted an effective increase in capacity for butadiene production. This may be offset by an increased cost for butylenes compared with butane.

Oxidative dehydrogenation processes, which may be applied to n-butane or n-butylenes for the production of butadiene, are considered in Chapter 9.

The trend towards heavier feedstocks for ethylene production in the U.S.A. is likely to lead to a situation, already evident in Europe, whereby amounts of butadiene, adequate for all needs, are available in the C_4 streams from the ethylene plants. This has already led to the virtual abandonment of these dehydrogenation plants in western Europe. It seems quite possible that the U.S.A. will follow suit by the early 1980's.

Acetic Acid from Oxidation of Naphtha

This is a relatively new process that accounts for nearly all the acetic acid produced in the United Kingdom.

The feedstock does not consist, strictly speaking, of paraffins, but as it comprises a paraffinic naphtha normally in the C_4–C_8 range, it may conveniently be dealt with under this heading.

The air oxidation proceeds in the liquid phase, typically at 160–200 °C and about 50 atm. The temperature is controlled so that an economic rate of reaction is maintained, without the production of excessive quantities of peroxide compounds.

The reaction mixture is flashed into a column where the lighter fractions (including unconverted hydrocarbons) are taken overhead, essentially for recycle. The bottom product is an aqueous acid mixture, which is worked up for the recovery of various products. The separation problems require the use of several columns. Water is

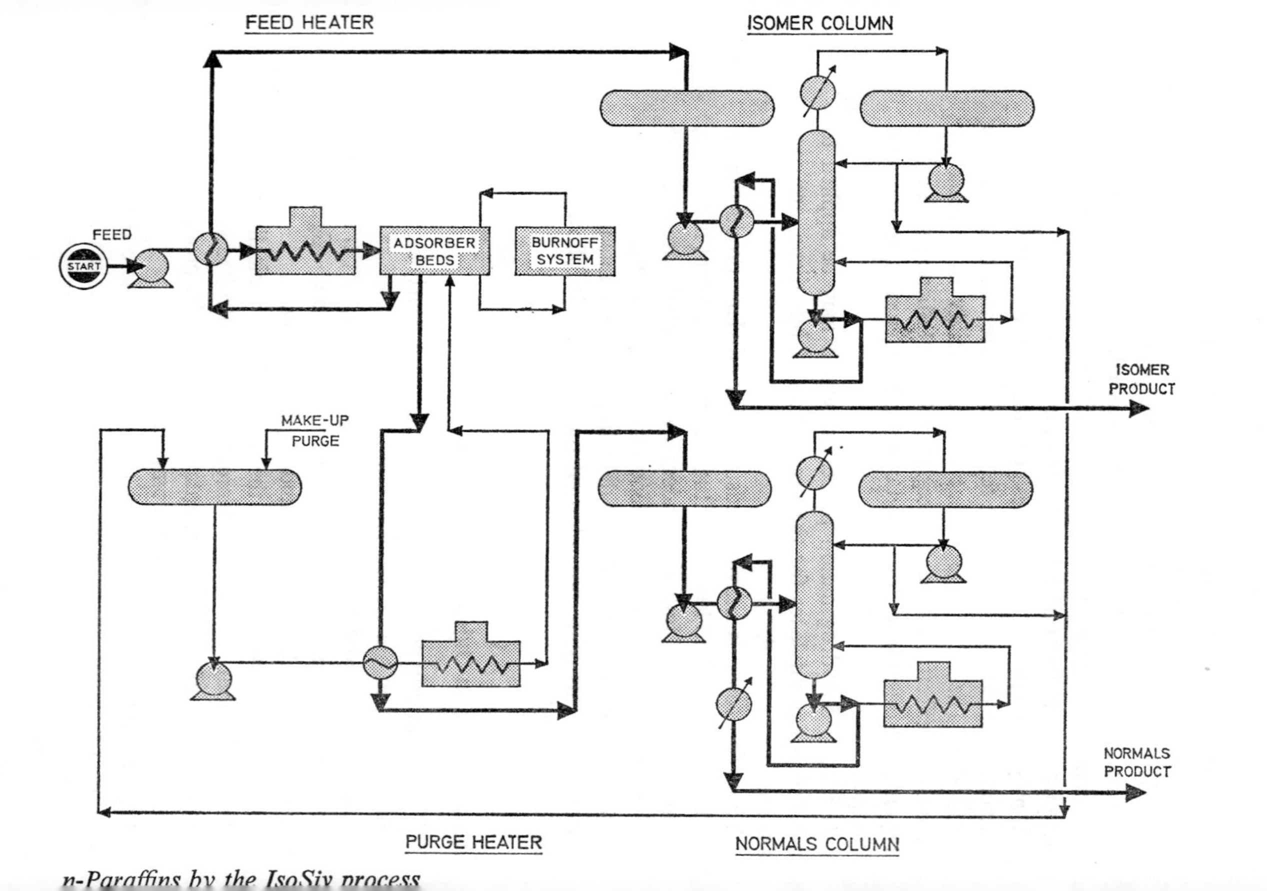

n-Paraffins by the IsoSiv process

eliminated by azeotropic distillation. Apart from acetic acid the main co-products are formic acid and propionic acid. Succinic acid is also recoverable from the acid mixture, and acetone may be extracted from the primary column overhead, before it passes to recycle.

A catalyst is not normally required but oil-soluble salts, such as the naphthenates of manganese, cobalt or copper, can be employed.

The process is now operated in several countries outside the United Kingdom, and it is probably the most economic process for acetic acid where naphtha is readily available and a satisfactory outlet exists for the co-product acids. In this connection it is interesting to observe the developing applications for formic and propionic acids in the preservation of agricultural produce, notably silage and a variety of cereals fed to animals.

Reactions of C_{10}–C_{18} Paraffins

The chemistry of the higher paraffins has encountered sharply increasing industrial interest since techniques were developed for the separation of the n-paraffins.

The principle of this is surprisingly simple. It stems from the development of molecular sieves. The dehydration of appropriate alkali-metal aluminosilicates (or synthetic zeolites) forms a crystalline structure with intracrystalline cavities, and the access to these is by minute pores of uniform size. This pore size is, within limits, controllable by adjustment of the chemical nature and processing of the molecular sieve. It is then possible to separate the molecules of straight chain paraffins from branched chain paraffins within the same boiling range by an effect similar to filtration.

The adsorption step may take place in the liquid or vapour phase. The n-paraffins are preferentially adsorbed on the molecular sieve and they are subsequently recovered by pressure release, or, more commonly, by washing them out with a more volatile n-paraffin (e.g. n-pentane). The n-paraffin product can then easily be separated from the volatile n-pentane, used as a wash liquor, by distillation. Since the higher paraffins are more strongly adsorbed, on the next adsorption cycle using fresh feed, the n-pentane is displaced from the sieve by the higher paraffin.

Separation of straight chain paraffins by means of the formation of a urea adduct is employed in this area, as well as for the treatment of wax used for wax cracking. The Nurex process uses urea in the solid phase, and the Edeleanu process operates in the liquid phase, with methylene dichloride as the liquid medium. In each case the separated adduct is decomposed under mild temperature conditions, and n-paraffins are recovered at 98 per cent purity.

The straight chain paraffins produced in this way, form the basis of the biodegradable detergents now used in many countries in an endeavour to avoid the problems of foaming in sewage works and sewage effluents, arising from the presence of detergent products which could not be properly treated by normal sewage processing.

There is a thoroughly traditional (and obsolete) process which started with the chlorination of a purified kerosene. The 'keryl chloride' was condensed with an excess of benzene at 50 °C using aluminium chloride as catalyst. This operation is an example of the process of alkylation, and the product was known as 'keryl benzene'. Sodium salts of sulphonated alkyl benzenes have been the main standby of the detergent business for many years. Keryl benzene was replaced in its turn by an alkyl benzene formed by the interaction of benzene and propylene tetramer, using hydrogen fluoride as the alkylation catalyst. These, however, were the products which were the least biodegradable, and which created the greatest nuisance in the sewage works and in the rivers.

With n-paraffins it is now common practice to revert to the old aluminium chloride alkylation technique. n-Paraffins in the C_{10}–C_{14} range are very gently chlorinated to give the monochloride. Typically the temperature may be 50–150 °C, and the conversion of the hydrocarbon is only taken to about 30 per cent, to avoid the formation of higher chlorinated products. The alkylation then proceeds as outlined for 'keryl chloride' above.

An alternative procedure is to take the n-paraffin in the C_{10}–C_{14} range and to subject it first to the mild chlorination described above, and then to remove a hydrogen chloride molecule to give the straight chain C_{10}–C_{14} olefin. Catalysts for this dehydrochlorination include silica gel/alkali halide at 250–400 °C and iron metal at about 300 °C. While this involves an additional processing step, it has the advantage that the alkylation of benzene using straight chain olefins can use the hydrogen fluoride catalyst employed in many existing production units. It is therefore possible to make greater use of existing facilities where these are available.

A further development has been the direct dehydrogenation process for producing straight chain olefins directly from n-paraffins. The dehydrogenation may be catalytically promoted using a platinum-based catalyst. Alternatively the operation may be a simple thermal cracking carried out on a similar basis to the well-established wax cracking process (see Chapter 3) using much the same conditions, and in certain circumstances some of the same equipment. Obviously, in view of the lower molecular weight of the paraffin feedstock, the conditions must be directed more towards dehydrogenation and less towards carbon chain rupture.

The straight chain olefins are also available from wax cracking (notably where the wax has been given a preliminary urea treatment to eliminate branched chain paraffins) and from the controlled polymerization of ethylene (see Chapter 7).

The chlorination processes using n-paraffins form hydrogen chloride as a by-product. This has traditionally posed a disposal problem, but the recently established oxychlorination processes (see Chapter 7) are amongst the factors which are rendering a hydrogen chloride by-product less troublesome.

Another obsolete detergent type, the German Mersolate or alkyl sulphonate, has also been resuscitated on the basis of the controlled sulphonation of appropriate n-paraffin fractions. These have been developed to the commercial stage both by Hoechst in Germany and by S.N.P.A. in France. The detergent products are admirably biodegradable but appear to have some processing disadvantages which are hindering their development.

The production of n-paraffins has become a major industry. The consumption in the U.S.A. was about 290 000 long tons in 1970 and this will rise to perhaps 335 000 long tons in 1975. The relatively slow growth stems from the fact that this usage depends on the production of detergents—not a particularly buoyant sector of the chemical industry. Of the total usage some 74 per cent finds application as linear detergent alkylate, and a further 21 per cent is for detergent alcohol production. These detergent alcohols may be of the primary or secondary type. In order to produce primary alcohols it is necessary first to dehydrogenate paraffins to olefins which are then subjected to the Oxo process (Chapter 11). Secondary alcohols may be produced directly from the paraffins by oxidation using boric acid (sometimes termed Bashkirov oxidation, Chapter 13). Another application of n-paraffins is in the form of chlorinated paraffins used as plasticizer extenders, notably for polyvinyl chloride.

Acetic acid from naphtha at Hull, England

Courtesy BP Chemicals International Ltd

Chapter 7

Ethylene Derivatives

Ethylene has proved to be one of the most versatile petroleum chemical raw materials. In scale of operation and variety of products it is the most important of the lower olefins. The tables below give an indication of the usage of ethylene both in tonnage and in terms of major derivatives.

ETHYLENE CONSUMPTION FOR CHEMICALS (IN '000 LONG TONS) IN U.S.A.

1960	*1965*	*1970*	*1971* (est.)	*1975* (est.)	*1980* (est.)
2 407	4 241	7 850	8 000	11 620	16 480

CONSUMPTION OF ETHYLENE ACCORDING TO END USE (AS %) IN U.S.A.

	1950	*1960*	*1965*	*1970*	*1975* (est.)	*1980* (est.)
Polyethylene	4	26	32	36	42	46
Ethylene oxide	30	29	24	21	19	17
Ethyl alcohol	32	21	17	8	6	4
Ethyl benzene	12	10	9	9	9	9
Halogen derivatives	11*	12	13	14	13	13
Miscellaneous	11	2	5	12†	11	11

* ethyl chloride only
† includes some ethylene alkylate for petroleum use

WEST EUROPEAN ETHYLENE PRODUCTION BY COUNTRIES

	1955	*1960*	*1965*	*1970*	*1975* (est.)	*1980* (est.)
Total west European production ('000 tons)	160	675	1 960	5 800	11 000	16 500
% in U.K.	66	37	27	16	13	
West Germany	22	34	35	34	28	
France	5	10	11	16	15	
Italy	7	13	18	14	14	
Benelux	—	} 6	5	13	20	
Others	—		4	7	10	

WEST EUROPEAN CONSUMPTION OF ETHYLENE BY PRODUCTS (AS %)

	1955	*1960*	*1965*	*1970*	*1975* (est.)
Polyethylene (low density)	} 30	37	38	38	36
Polyethylene (high density)		7	12	12	13
Ethylene dichloride	*	5	6	16	20
Ethyl benzene	*	9	8	8	8
Ethylene oxide	48	26	19	15	13
Ethyl alcohol	16	10	7	3	3
Acetaldehyde	*	4	8	4	3
Miscellaneous	6	2	2	4	4

* included in miscellaneous

Capacity elsewhere for ethylene production is also substantial. Japan is an outstanding example with an annual capacity of about 4 million tons. Canadian capacity is over 500 000 tons per year, and in 1970 Canada consumed about 404 000 long tons of ethylene.

Growth in ethylene usage will not maintain the spectacular level of the past, but will remain very substantial. In western Europe a growth of some 14 per cent a year is envisaged, and growth in the U.S.A. is expected to be 8–10 per cent a year.

The ramifications of ethylene chemistry are such as to make it necessary to deal with the derivatives under separate headings, which will themselves require subdivision in certain cases.

Ethyl Alcohol

Production of Ethyl Alcohol

In many countries, the great bulk of ethyl alcohol is still obtained by fermentation processes. Where there is an unquestioned surplus of fermentable material (molasses, grain, starch, wine residues, etc.) of negligible value, the fermentation process remains competitive. Once the element of doubt creeps in, the value of the supposedly surplus product becomes unpredictable. An example of this has been molasses from some sugar-producing areas, which has in the past represented an important raw material for ethyl alcohol. Once conditions of low and stable prices for a raw material are in doubt, the whole economic basis of the fermentation process for ethyl alcohol comes into question. The proportion of ethyl alcohol made synthetically, in both the U.S.A. and the United Kingdom, varies to some extent according to the short term supply position of fermentable materials, and some occasional shortages of synthesis capacity.

In the U.S.A. fermentation alcohol may represent up to about 10 per cent of total production and in the United Kingdom up to about 15 per cent.

In the U.S.A. the production of industrial ethyl alcohol in 1970 was about 890 000 long tons. This figure shows a decline compared with some recent years, and reflects the inroads made into the major ethyl alcohol market of acetaldehyde production. Fermentation alcohol only accounted for 3 per cent of the total.

In the United Kingdom the total production of industrial ethyl alcohol was 128 300 long tons in 1970 and about 5 per cent of this was from fermentation.

There are two main processes for the synthesis of ethyl alcohol from ethylene. The earliest to be developed was a process of the concentration-dilution type. This involves the initial absorption of ethylene into 96 per cent sulphuric acid at 55–75 °C and about 24 atm. The pressure, in theory, is varied according to the purity of the ethylene stream, but in practice the ethylene stream used today is rarely of less than about 96 per cent purity. It should be free from higher olefins. The liquor resulting from the ethylene absorption is a mixture of monoethyl and diethyl sulphates. The mixed esters are pumped to a hydrolyser, together with an appropriate volume of water. Hydrolysis takes place using steam heating. The mixture then contains ethyl alcohol, dilute sulphuric acid and small amounts of ethyl ether (5–10 per cent of total production) and other by-products. This passes to a stripping column, where steam carries the volatile components overhead, allowing dilute sulphuric acid to be drawn off at the bottom.

$$CH_2{=}CH_2 + H_2SO_4 \rightarrow \underset{\text{monoethyl sulphate}}{C_2H_5OSO_2OH}$$

$$2CH_2{=}CH_2 + H_2SO_4 \rightarrow \underset{\text{diethyl sulphate}}{C_2H_5OSO_2OC_2H_5}$$

$$C_2H_5OSO_2OH + C_2H_5OSO_2OC_2H_5 + 3H_2O \rightarrow 3C_2H_5OH + 2H_2SO_4$$

The acid is concentrated by vacuum evaporation and re-used. The crude alcohol vapour from the top of the stripping column passes to a scrubber to remove traces of acidity. Ethyl ether is removed by steam in a distillation column, and the alcohol passes to a fractionating column, from which an azeotrope with water containing 95.6 per cent volume alcohol is taken overhead.

Where anhydrous alcohol is required, the 95.6 per cent azeotrope may be distilled with an entraining agent such as benzene. The benzene forms a ternary azeotrope with alcohol and water boiling at 69.7 °C and a binary azeotrope with alcohol boiling at 72.5 °C. These are separated from the anhydrous alcohol boiling at 78.4 °C.

A more recent synthesis process is designed to eliminate the use of sulphuric acid. This is the direct hydration process; it typifies the modern trend towards the most direct possible synthesis route.

$$CH_2{=}CH_2 + H_2O \rightleftharpoons C_2H_5OH$$

In this case the olefin stream will be pure ethylene containing traces of methane, ethane, acetylene and higher olefins. It is perhaps necessary to stress that in a recycle process of this kind it is always desirable to minimize impurities (even if they are merely diluents) in the feed. Such impurities will accumulate in the recycle streams and they require a systematic purge (or exhaustion to atmosphere) to prevent such a build-up taking place. The purge always occasions some loss of valuable raw material. The ethylene is compressed to 68 atmospheres, and mixed with a stream of water. The mixture is reacted in the vapour phase at 300 °C over a catalyst of phosphoric acid on a porous inert support such as diatomaceous earth or Celite. The choice of conditions is dictated by the familiar conflict of economic requirements. As the temperature increases, the rate of reaction increases, but the equilibrium concentration of ethyl alcohol decreases. The ethylene conversion per pass is only 4–5 per cent, so that a considerable proportion of the ethylene is recycled. It is theoretically possible to achieve an ethylene conversion per pass up to 20 per cent by using a high mol ratio of water to ethylene, and a high pressure, but the economics of such an operation are unfavourable.

The product stream leaving the reactor is partly condensed, scrubbed with sodium hydroxide to remove traces of acidity, and then treated by a variety of separation processes to recover all the alcohol in the condensed liquids. The residual gas stream requires to be purged (i.e. given a bleed into the atmosphere or vented back to the ethylene plant) in order to prevent the accumulation of inert components (methane and ethane) in the recycle gas stream. In the alcohol/water mixture there may be traces of acetaldehyde from the hydration of any acetylene present. The separation process applied to this mixture therefore probably includes a vapour phase hydrogenation with a nickel catalyst. The alcohol is concentrated as before.

This process, like many others, has been subjected to continuous study and modification. Attention has been devoted particularly to the specific form of the catalyst, and to the overall distillation procedure. Improvements have recently been made both in the utility requirements for this distillation and in the purity of the final product.

It is an interesting sidelight on the variability of economic factors, that in one or two areas it is still possible to justify the reverse process of obtaining ethylene by dehydration of ethyl alcohol.

Applications of Ethyl Alcohol

The usage of ethyl alcohol in the U.S.A. in 1970 was as follows:

Acetaldehyde production	29%
Miscellaneous chemical synthesis	28%
Solvent uses	41%
Miscellaneous	2%

It was not until 1967 that solvent applications caught up with the traditional acetaldehyde production as the major ethyl alcohol outlet. Other processes for acetaldehyde production have long been used. The acetylene-based production is now largely obsolete, and production from paraffins has never been of major importance. During the 1960's the direct oxidation of ethylene to acetaldehyde became an important process (the Wacker process—see later in this chapter). This development, coupled with alternative processes for making acetic acid, has sharply diminished the production of acetaldehyde from ethyl alcohol in the U.S.A. as elsewhere. Such processes are still of importance, however, and it has been suggested that improvements in the economics of ethyl alcohol manufacture may improve their competitive position.

The production of acetaldehyde from ethyl alcohol may take the form of an oxidation or a dehydrogenation.

The production of acetaldehyde by oxidation of ethyl alcohol with air

$$2C_2H_5OH + O_2 \longrightarrow 2CH_3CHO + 2H_2O$$

involves passing alcohol vapours and preheated air at 450 °C and 3 atmospheres pressure over a silver gauze catalyst. With the appropriate proportion of air (2 mols per mol alcohol) the exothermic heat of oxidation will balance the endothermic dehydrogenation reaction

$$C_2H_5OH \rightarrow CH_3CHO + H_2$$

so that no external application of heat is necessary.

The dehydrogenation process, which is more common, requires a chromium-oxide-activated copper catalyst. The temperature is maintained at 270–300 °C. The fairly pure hydrogen stream can have useful applications.

With either process the acetaldehyde is separated by distillation from the unreacted alcohol, which is recycled.

The main consumption of acetaldehyde is in the production of acetic acid and acetic anhydride. The liquid phase oxidation of acetaldehyde to acetic acid is carried out by the use of air or oxygen in the presence of manganous acetate, or cobalt acetate,

$$2CH_3CHO + O_2 \rightarrow 2CH_3COOH$$

Reaction conditions using oxygen are 70–80 °C and a pressure sufficient to keep the acetaldehyde liquid. Using air, the conditions quoted have been 40–65 °C and a pressure of about 5 atmospheres.

A modification of this reaction to allow for the simultaneous production of acetic acid and acetic anhydride is achieved by use of a mixture of copper and cobalt acetates as the catalyst. It is common to add a substantial proportion of ethyl acetate as a diluent. This has the effect of increasing the ratio of acetic anhydride to acetic acid in the product mixture. The reaction conditions may typically be 50–70 °C and 5 atmospheres pressure.

$$\underset{\text{acetaldehyde}}{CH_3CHO} + O_2 \rightarrow \underset{\text{peracetic acid}}{CH_3COOOH}$$

$$\underset{\text{paracetic acid}}{CH_3COOOH} + \underset{\text{acetaldehyde}}{CH_3CHO} \rightarrow \underset{\text{acetic acid}}{2CH_3COOH}$$

$$CH_3COOOH + CH_3CHO \rightarrow \underset{\text{acetic anhydride}}{(CH_3CO)_2O} + H_2O$$

An incidental feature of this process is that variations on it may be used specifically for the production of peracetic acid. The Union Carbide operation is a liquid-phase oxidation of acetaldehyde said to proceed via the intermediate product acetaldehyde monoperacetate, which decomposes with heat to form peracetic acid. This acid is used in the Union Carbide process to caprolactam (Chapter 14). A vapour phase oxidation of acetaldehyde to peracetic acid has been operated in this country. Peracetic acid is a valuable oxidizing agent.

The most favoured process for the manufacture of acetic anhydride starts with the catalytic pyrolysis of acetic acid. The acid is cracked at 700 °C and under reduced pressure in the presence of triethyl phosphate. The gases from the furnace pass to a condenser and aqueous acetic acid is drawn off at the bottom. At the top nearly pure ketene passes out in the gaseous form. The ketene is immediately reacted with further acetic acid to give acetic anhydride.

$$\underset{\text{acetic acid}}{CH_3COOH} \rightarrow \underset{\text{ketene}}{CH_2{=}C{=}O} + H_2O$$

$$CH_3COOH + CH_2{=}C{=}O \rightarrow (CH_3CO)_2O$$

While perhaps 95 per cent of ketene is used to make acetic anhydride there are other interesting applications. With ethyl alcohol it will form ethyl acetoacetate, a dyestuff intermediate. With paraformaldehyde ketene reacts to form 2-propiolactone, an intermediate in a commercial route to acrylic acid.

The alternative route to acetic anhydride with ethylidene diacetate as intermediate must now be regarded as obsolete. The reactions are:

$$\underset{\text{acetic acid}}{2CH_3COOH} + \underset{\text{acetylene}}{HC{\equiv}CH} \rightarrow \underset{\text{ethylidene diacetate}}{CH_3CH(OCOCH_3)_2}$$

$$\underset{\text{ethylidene diacetate}}{CH_3CH(OCOCH_3)_2} \rightarrow \underset{\text{acetic anhydride}}{(CH_3CO)_2O} + \underset{\text{acetaldehyde}}{CH_3CHO}$$

It is an interesting reflection on the confused state of the acetyl product manufacturing picture that it is perfectly feasible to move some of these reactions in the opposite direction. Starting with acetic anhydride and acetaldehyde, and proceeding through ethylidene diacetate as an intermediate product, one may produce vinyl acetate together with acetic acid. In such a case the acetic acid would be recycled to produce more acetic anhydride. This operation in acetyl chemistry was essentially based on the acetyl products of paraffin oxidation, but it can scarcely face the competition of ethylene-based vinyl acetate.

The consumption of acetaldehyde in the U.S.A. amounted to about 715 000 long tons in 1970 and the figure is expected to rise to 800 000 tons in 1974. The breakdown of usage for 1970 was:

Acetic acid and anhydride	45%
n-Butyl alcohol	19%
2-Ethylhexanol	17%
Miscellaneous	19%

Some detailed acetaldehyde figures are also available from Japan. The total demand there is scheduled to rise from 535 000 tons in 1970 to 723 000 tons in 1975. The breakdown of this usage has been estimated as follows:

	1970	*1975*
Acetic acid	56%	59%
Ethyl acetate	15%	14%
n-Butyl alcohol	16%	17%
Pentaerythritol	2%	2%
Miscellaneous	11%	8%

A recent estimation put the proportion of acetaldehyde capacity in the U.S.A. accountable to direct oxidation of ethylene as 56 per cent of the total; 32 per cent was based on ethyl alcohol, 11 per cent on paraffin oxidation, and 1 per cent was of by-product origin.

Figures for acetic acid production are inclined to be variable because of confusion arising from the recycling of acid in cellulose acetate manufacture. The demand figure for the U.S.A. in 1970 was around 815 000 long tons and a usage pattern has been quoted as follows:

Cellulose acetate	44%
Vinyl acetate	31%
Other acetate esters	14%
Monochloroacetic acid	3%
Miscellaneous	8%

A steady growth is forecast in U.S. acetic acid consumption to about 1 200 000 tons in 1975. The main area of growth is in the vinyl acetate application.

The production of acetic anhydride has not been growing very fast. In the U.S.A., for example, the 1966 production was about 715 000 tons, and this rose to about 735 000 tons in 1970. It has, however, been suggested that U.S. demand might reach 800 000 tons by 1973. A consistent 90 per cent of this usage is for cellulose esters (mostly acetate, but including acetate-butyrate and acetate-propionate). About 5 per cent goes into vinyl acetate, and another 5 per cent to miscellaneous uses, of which perhaps the most important is aspirin.

The second most important application for acetaldehyde is probably the complex synthesis of n-butyl alcohol.

Acetaldehyde may be dimerized at 20 °C in the presence of dilute alkali to form acetaldol.

$$\underset{\text{acetaldehyde}}{CH_3CHO + CH_3CHO} \rightarrow \underset{\text{acetaldol}}{CH_3CH(OH)CH_2CHO}$$

The aldolization is allowed to proceed to 60 per cent conversion. The catalyst is neutralized, a slight excess of acid added, and acetaldol dehydrated in a two-column distillation unit.

$$\underset{\text{acetaldol}}{CH_3CH(OH)CH_2CHO} \rightarrow \underset{\text{crotonaldehyde}}{CH_3CH{=}CHCHO} + H_2O$$

$$\underset{\text{crotonaldehyde}}{CH_3CH{=}CHCHO} + 2H_2 \rightarrow \underset{\text{n-butyl alcohol}}{CH_3CH_2CH_2CH_2OH}$$

n-Butyl alcohol is obtained by the vapour phase hydrogenation of crotonaldehyde, using a nickel-chromium catalyst at 180 °C and slightly elevated pressure. Advantage may be taken of the fact that where the acetaldehyde is made by dehydrogenation of ethyl alcohol, the hydrogen available is exactly appropriate to the hydrogen requirement for crotonaldehyde hydrogenation. In spite of this there is increasing preference for the Oxo route to n-butyl alcohol starting with propylene (Chapter 8).

n-Butyl alcohol is used as a solvent, either directly or in the form of the acetate ester. It also has important applications as a chemical intermediate in the production of glycol ethers and plasticizers.

Ethyl alcohol is converted to ethyl chloride by heating it with hydrogen chloride in the presence of zinc chloride at 145 °C and slightly elevated pressure. This was the original route to ethyl chloride. It is still used, but only on a minor scale.

$$\underset{\text{ethyl alcohol}}{C_2H_5OH} + HCl \rightarrow \underset{\text{ethyl chloride}}{C_2H_5Cl} + H_2O$$

The esterification of ethyl alcohol and acetic acid, in the presence of a catalyst such as sulphuric acid, produces ethyl acetate. The equilibrium conversion is about 67 per cent. The reaction may be forced to completion by removing the water formed and employing one reactant in excess.

$$\underset{\text{ethyl alcohol}}{C_2H_5OH} + \underset{\text{acetic acid}}{CH_3COOH} \rightleftharpoons \underset{\text{ethyl acetate}}{CH_3COOC_2H_5} + H_2O$$

Whilst the esterification route to ethyl acetate is the most common, it is possible to employ the Tischtschenko synthesis of esters from aldehydes (a variant of the Canizzaro reaction). Acetaldehyde in the presence of some ethyl acetate and ethyl alcohol will condense to ethyl acetate using aluminium ethoxide as catalyst at a temperature of 0 °C.

$$\underset{\text{acetaldehyde}}{2CH_3CHO} \rightarrow \underset{\text{ethyl acetate}}{CH_3COOC_2H_5}$$

Ethyl acetate is an important solvent. Production in the U.S.A. has risen from 53 000 tons in 1966 to about 68 000 long tons in 1970. Future growth will follow much the same pattern—a figure of 4 per cent growth per annum has been forecast. Almost the whole usage of ethyl acetate (97 per cent) is for solvent applications—70 per cent for surface coatings, 10 per cent for plastics and 17 per cent for miscellaneous solvent uses. The remaining usage is for chemical synthesis.

There are six producers of ethyl acetate in the U.S.A., but only two in the United Kingdom. The Tischtschenko synthesis is not operated in this country, and may be regarded as basically obsolete, but it is still in operation in a number of places.

As was indicated earlier, the whole field of 'acetyl chemistry' is in a fluid state of development arising from new technology in acetaldehyde and acetic acid production. This is likely to continue to affect the applications of ethyl alcohol for chemical synthesis. On the other hand the solvent applications of ethyl alcohol, now the most important single outlet, continue to expand. These take effect in such preparations as lacquers, varnishes, cellulose nitrate, polishes and cosmetics.

Mention should perhaps be made of the historical application of ethyl alcohol to make butadiene. This played a significant part in the development of synthetic rubber during the second world war. Today this process is limited to countries where other considerations may take precedence over economics. The reactions are outlined as follows:

$$\underset{\substack{\text{ethyl}\\\text{alcohol}}}{2C_2H_5OH} \xrightarrow{-2H_2} \underset{\text{acetaldehyde}}{2CH_3CHO} \longrightarrow \underset{\text{crotonaldehyde}}{CH_3CH{=}CHCHO} + H_2O$$

$$\underset{\substack{\text{ethyl}\\\text{alcohol}}}{C_2H_5OH} + \underset{\text{crotonaldehyde}}{CH_3CH{=}CHCHO} \longrightarrow \underset{\text{butadiene}}{CH_2{=}CHCH{=}CH_2} + \underset{\text{acetaldehyde}}{CH_3CHO} + H_2O$$

Ethylene Oxide

Production of Ethylene Oxide

Today nearly all ethylene oxide is produced by the direct oxidation of ethylene. Historically, the first major development was an indirect oxidation, involving ethylene chlorohydrin as intermediate. This process is still operated, but only on a modest scale, and is likely to be phased out altogether within a few years. The reactions are as follows:

$$Cl_2 + H_2O \rightleftharpoons HOCl + HCl$$

$$CH_2{=}CH_2 + Cl_2 + H_2O \rightarrow \underset{\text{ethylene chlorohydrin}}{CH_2OH{-}CH_2Cl} + HCl$$

$$2\underset{\text{ethylene chlorohydrin}}{CH_2OH{-}CH_2Cl} + Ca(OH)_2 \rightarrow 2\underset{\text{ethylene oxide}}{CH_2{-}CH_2(O)} + CaCl_2 + 2H_2O$$

Although such a process offers quite a good yield of ethylene oxide on ethylene, the economic objection to using chlorine in the reaction when it is ultimately converted into a waste product—calcium chloride—proved overwhelming.

Direct oxidation processes began to gain favour during the 1950's. The most effective catalyst was found to be silver. The oxidation may be carried out using air or oxygen. Economies can be effected in the oxidation unit by using oxygen, but these are offset by the cost of an air separation unit. Both forms of the process are used, and the balance of advantage will depend upon local conditions, the scale of the operation (a very large scale has tended to favour the oxygen route), and the value of the co-product nitrogen.

The two main reactions which occur are:

$$\underset{\text{ethylene}}{CH_2{=}CH_2} + \tfrac{1}{2}O_2 \rightarrow \underset{\text{ethylene oxide}}{CH_2{-}CH_2(O)}$$

$$CH_2{=}CH_2 + 3O_2 \rightarrow 2CO_2 + 2H_2O$$

The ethylene stream should be fairly pure (95 per cent volume) or side reactions may interfere. The problem here is not one of recycling impurities (which must be purged) but of maximizing the yield of the desired product. Ethylene streams are today normally provided in high purity. The temperature is commonly 260–290 °C for air oxidation, or nearer 230 °C where oxygen is used. It is possible to operate the process at atmospheric pressure, but now pressures of 10–30 atm. are commonly used.

The ethylene oxide is isolated by scrubbing the reaction gases with water at atmospheric or higher pressure. The ethylene oxide is subsequently stripped from solution and purified by distillation.

Strict control of the reaction conditions is necessary to minimize loss of ethylene as carbon dioxide in accordance with the second of the equations above.

Very little ethylene oxide is now made by the chlorohydrin route. The usage of ethylene per ton of ethylene oxide is generally marginally over 1 ton, whereas only about 0.9 ton of ethylene was required when the chlorohydrin route was followed. The chlorohydrin plants also involved slightly lower capital cost than the direct oxidation plants. Ethylene oxide by the chlorohydrin route requires 2 tons of chlorine and 2 tons of lime per ton of product made. Economic trends of the 1960's, including the reduction of ethylene price, the increase of chlorine price, and massive increases in scale, all told against the chlorohydrin route.

While all new ethylene oxide plants employ direct oxidation, the old chlorohydrin plants have in many cases been switched to propylene oxide manufacture, where the chlorohydrin process is still pre-eminent.

Ethylene oxide is produced on a vast scale. The production in the U.S.A. was about 1 690 000 long tons in 1970 and the forecast production for 1974 is almost 2 million tons. In 1970 the production in the United Kingdom was 174 600 long tons.

Applications of Ethylene Oxide

The relative importance of various applications of ethylene oxide is indicated in the following recent table for usage:

	U.S.A.	*Japan*
Ethylene glycol	56%	66%
Nonionic surface active agents	17%	17%
Polyglycols	7%	3%
Ethanolamines	8%	7%
Glycol ethers } Others }	12%	4%
Others		3%

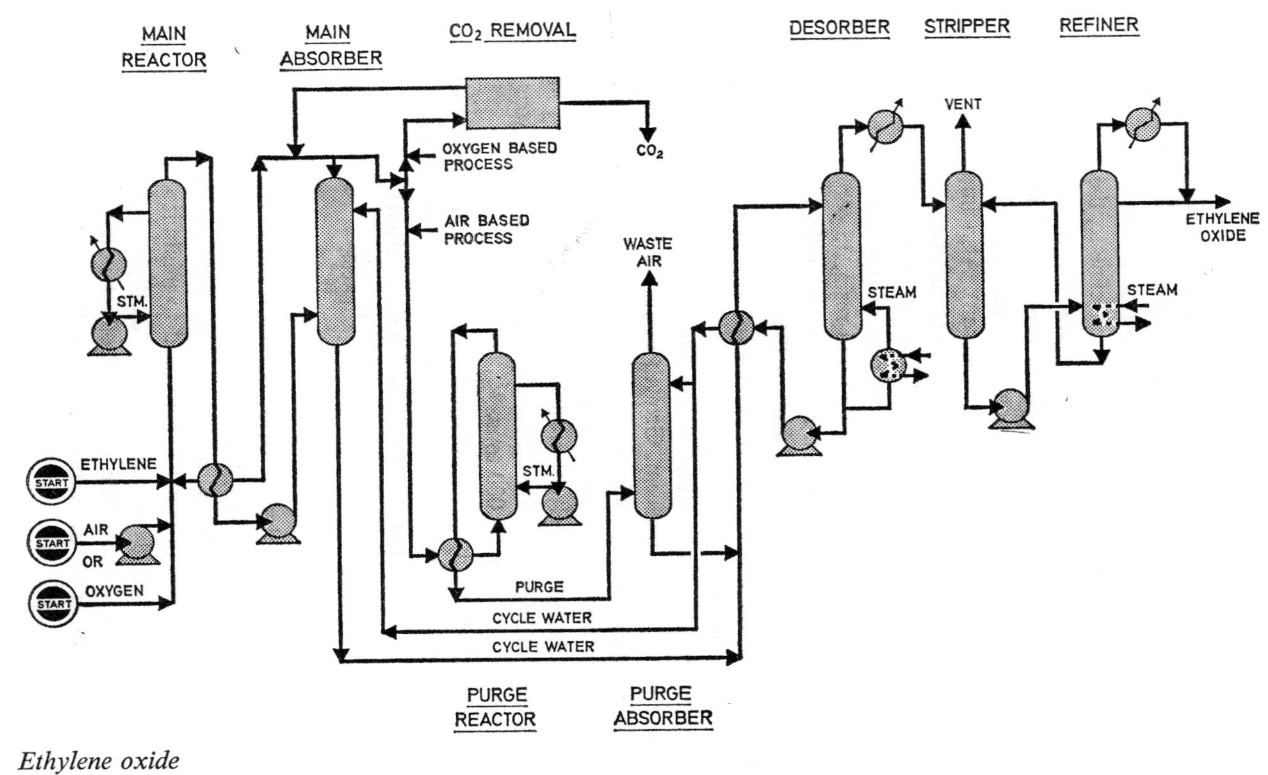

Ethylene oxide

(From *Hydrocarbon Processing*, Nov. 1971)

In almost every major industrial country it is now true to say that the main application for ethylene oxide is in the production of ethylene glycol for anti-freeze in car radiators. Whilst there are obvious differences in the degree of motorization, the size of car used, the popularity of air-cooled engines, and climate, the proportional figures are tending to assume some degree of similarity.

Other markets for ethylene glycol are also of importance, the most significant development being the increasing scale of production of polyester fibres (e.g. Terylene, Dacron). Also important is the production of regenerated cellulose film (Cellophane).

Ethylene glycol usage in the U.S.A. in 1970 approximated to 1.25 million long tons. Some 59 per cent of this was for anti-freeze, 17 per cent for polyester fibre, film and resin, 13 per cent for other uses and 11 per cent for export. The export figure had been declining, but 1970 was an exceptional year. In 1966 over 70 per cent of U.S. ethylene glycol usage was for anti-freeze and only 6 per cent for polyester fibre, film and resin.

Elsewhere the proportion of ethylene glycol moving into anti-freeze is near to 70 per cent. Growth in anti-freeze markets is only about 3.5 per cent a year in the U.S.A., but it is 7 per cent a year in western Europe and 10 per cent in Japan.

In Canada ethylene glycol production was quoted as being 74 000 long tons in 1970, and the Canadian consumption in that year was marginally higher at 78 000 tons.

Some ethylene glycol is obtained as a by-product in the direct oxidation process for ethylene oxide but most is made by hydration.

$$\underset{\text{ethylene oxide}}{\overset{CH_2—CH_2}{\diagdown_{O}\diagup}} + H_2O \rightarrow \underset{\text{ethylene glycol}}{HOCH_2CH_2OH}$$

The hydration of ethylene oxide is carried out in the liquid phase (sometimes with an acid catalyst) at atmospheric pressure and 50–70 °C, or at about 195 °C under 13 atm. pressure, but without a catalyst. An excess of water is employed to minimize the inevitable formation of polyglycols (principally diethylene glycol and triethylene glycol). The crude reactor product is concentrated to 70 per cent glycols by evaporation. The subsequent distillation procedure is complex, since water has to be eliminated, and the individual glycols separated in a pure and marketable condition.

An alternative process for ethylene glycol, which may now be regarded as obsolescent, incorporates the following reactions:

$$\underset{\text{formaldehyde}}{CH_2O} + CO + H_2O \rightarrow \underset{\text{glycollic acid}}{CH_2OHCOOH}$$

$$CH_2OHCOOH + \underset{\text{methyl alcohol}}{CH_3OH} \rightarrow \underset{\text{methyl glycollate}}{CH_2OHCOOCH_3} + H_2O$$

$$CH_2OHCOOCH_3 + 2H_2 \rightarrow \underset{\text{ethylene glycol}}{CH_2OHCH_2OH} + CH_3OH$$

The co-products of ethylene oxide hydration, diethylene glycol and triethylene glycol, are both industrial products of significance in their own right. Diethylene glycol, which was consumed to the extent of nearly 100 000 long tons in the U.S.A. in 1970, has a variety of applications:

Polyurethane and unsaturated polyester resins	30%
Triethylene glycol	13%
Textile agents	12%
Solvent extraction (Udex process)	7%
Natural gas dehydration	7%
Plasticizers and surface active agents	7%
Exports	10%
Miscellaneous	14%

Triethylene glycol usage in the U.S.A. was around 38 000 long tons in 1970, and once again the applications are most varied:

Natural gas dehydration	28%
Humectant	19%
Solvent (for printing inks etc.)	15%
Exports	10%
Vinyl plasticizer	9%
Udex process extraction solvent	8%
Polyester and urethane resins	6%
Miscellaneous	5%

Polyethylene glycols are made by continuing the reaction between ethylene oxide and ethylene glycol or diethylene glycol. The reaction takes place at 120–150 °C and 3 atmospheres pressure, with the assistance of an alkaline catalyst such as sodium hydroxide. Alternatively ethylene oxide may be passed into a small amount of preformed polyglycol. According to molecular weight, the polyethylene glycols are liquids or waxy solids. The lower molecular weight products are brake fluid components, and find important applications as plasticizers, and humectants in the production of transparent adhesive tapes. The medium and higher molecular weight products are lubricants in the processing of textile fibres, and components of pharmaceutical and cosmetic formulations. The production of polyglycols in the U.S.A. has not quite kept pace with the growth of ethylene oxide as a whole and is of the order of 25 000 tons per year.

Ethylene oxide will react with alcohols, in the presence of excess alcohol and a catalyst, which may be sodium hydroxide or boron trifluoride, according to the alcohol being treated. The conditions also vary according to the alcohol, but with ethyl alcohol, the reaction temperature may be 170–190 °C, and the pressure 13–16 atmospheres. The product in this case is ethylene glycol monoethyl ether. The range of products is described generically as glycol ethers.

The monoethers are the most important products, but the reaction normally provides a proportion of di- and triethers, which are also separated for a range of industrial uses. Several related products are made on the commercial scale, but the most important are the methyl, ethyl and n-butyl monoethers of ethylene glycol.

$$C_2H_5OH + \underset{\diagdown\,O\,\diagup}{CH_2{-}CH_2} \rightarrow \underset{\text{ethylene glycol monoethyl ether}}{C_2H_5OCH_2CH_2OH}$$

The free hydroxyl group can be esterified. Some of the acetate esters are also of considerable importance in the surface-coatings field.

$$\underset{\text{ethylene glycol monoethyl ether}}{C_2H_5OCH_2CH_2OH} + \underset{\text{acetic acid}}{CH_3COOH} \rightarrow \underset{\text{ethylene glycol monoethyl ether acetate}}{C_2H_5OCH_2CH_2OCOCH_3}$$

Other applications for this range of products include brake fluids (as diluents for viscosity control), military jet fuel additives (to combat both ice formation and bacterial corrosion), plasticizers, cleaning compounds, printing inks and various formulated products.

Nonionic surface active agents account for 17 per cent of U.S. ethylene oxide and a slightly smaller proportion—perhaps 15 per cent—of western European ethylene oxide. The question of biodegradability which first loomed large in the field of anionic detergents (Chapter 17), has more recently had an important influence on the chemical structure of nonionic detergents. It is now generally accepted that the alkyl phenol ethoxylates are biologically 'hard' whereas the alcohol ethoxylates are relatively readily biodegradable.

Nonionic detergents are growing faster than detergents as a whole for a variety of reasons, but notably as a result of their suitability for the 'low-foam' formulations needed to an increasing extent in automatic washing machines. Nonionic detergents made from ethylene oxide can also be tailored to suit liquid detergent formulations. Here their ready solubility is an advantage. Sulphated ethoxylates of higher alcohols are particularly suitable products for liquid household detergents.

The condensation products of alkyl phenols with ethylene oxide use particularly nonyl phenol, and also octyl and dodecyl phenols, as raw materials. The alkyl phenol is typically condensed with 8–12 mols ethylene oxide per mol. Ethylene oxide is added to the alkyl phenol, using sodium acetate or sodium hydroxide as a catalyst, at such a rate as will keep the temperature at about 200 °C. The pressure is kept at 2–2.5 atmospheres during the addition.

Consumption of nonionic surface active agents worldwide in 1970 amounted to 916 000 long tons. The U.S. contribution to this total was 472 000 long tons; of this total ethoxylated alkyl phenols amounted to 133 000, and ethoxylated alcohols 197 000 tons. Forecasts for 1975 indicated the following:

	Nonionic detergent consumption	*Ethoxylated alkyl phenols*	*Ethoxylated alcohols*
U.S.A.	786 000 long tons	217 000 long tons	374 000 long tons
Europe	443 000 long tons	138 000 long tons	207 000 long tons

The alcohols used for the production are commonly either in the C_{12}–C_{15} range (when they may be condensed with 3–10 mols ethylene oxide per mol alcohol) or C_{15}–C_{18} products (which may be typically condensed with 10–25 mols ethylene oxide per mol alcohol). The C_{12}–C_{15} alcohol ethoxylates are suitable for subsequent sulphation and neutralization to make liquid household detergents. The C_{15}–C_{18} alcohol ethoxylate provides a low-foaming detergent formulation.

Where it is desired to reproduce from alcohols an equivalent to the conventional alkyl phenol derivatives, a C_{10}–C_{15} alcohol fraction may be condensed with 6–8 mols ethylene oxide per mol alcohol.

It is only in the past few years that the alcohol ethoxylates have overtaken the alkyl phenol ethoxylates in scale of production. For the future it is expected that the alcohol derivatives will grow much faster than those from alkyl phenols. This trend stems from the factor of biodegradability mentioned above.

Ethylene oxide will react with ammonia to give a mixture of ethanolamines.

$$\underset{\text{(ethylene oxide: } CH_2-CH_2 \text{ bridged by O)}}{CH_2\!-\!CH_2} + NH_3 \rightarrow \underset{\text{monoethanolamine}}{NH_2CH_2CH_2OH}$$

$$NH_2CH_2CH_2OH + \underset{\text{(bridged by O)}}{CH_2\!-\!CH_2} \rightarrow \underset{\text{diethanolamine}}{NH(CH_2CH_2OH)_2}$$

$$NH(CH_2CH_2OH)_2 + \underset{\text{(bridged by O)}}{CH_2\!-\!CH_2} \rightarrow \underset{\text{triethanolamine}}{N(CH_2CH_2OH)_3}$$

The exothermic reaction takes place under conditions which may be varied according to the relative requirement of the individual amines. An excess of ammonia is used. The greater the excess, the higher the proportion of monoethanolamine in the final mixture. Temperature and pressure also have some influence on the product ratio. Pressure may vary from atmospheric to 100 atmospheres, and temperature from 30 to 270 °C. Using 30–40 °C and a pressure of 1.5 atmospheres, the reaction of ammonia and ethylene oxide in a 10 : 1 mol ratio will yield 75 per cent monoethanolamine, 21 per cent diethanolamine and 4 per cent triethanolamine. Using equimolar proportions under the same conditions, the respective proportions become 12, 23 and 65 per cent. Where triethanolamine is the desired product, and low ammonia/ethylene oxide mol ratios are consequently used, some carbon dioxide may be required as a diluent, to suppress ether formation. Alternatively, the mono- or diamine can be recycled to repress its proportion in the product mixture. The usual separation by distillation in a multicolumn complex is carried out. Ammonia is stripped off the amine mixture and recycled to the reaction.

Alternative conditions which have been reported for the basic reaction are a temperature of rather under 150 °C, a reaction time of 1–2 minutes and a pressure of about 65 atm.

Ethanolamine consumption in the U.S.A. reached about 120 000 long tons in 1970. Forecast usage in 1975 is 154 000 long tons. Some degree of flexibility in the manufacturing process has proved necessary since the relative requirements of different amines have proved somewhat variable.

Monoethanolamine has an important application as a scrubbing liquid for removing acid gases (e.g. hydrogen sulphide, carbon dioxide) from gas streams. It is a chemical intermediate for alkanolamides used as foam boosters in detergent formulations, for a variety of emulsifying agents, and for piperazine.

Diethanolamine is also used for gas scrubbing and the production of detergent additives. A specific application of diethanolamine is its dehydration to morpholine.

Triethanolamine has a major application in cosmetic formulations and textile auxiliaries. The mildness of its alkalinity accounts for much of its desirability in these applications.

The combined usage pattern has been quoted for 1970 as follows:

Detergents (including speciality surface active agents)	37%
Gas conditioning and petroleum use	20%
Exports	17%
Morpholine	7%
Miscellaneous	19%

The gas conditioning application is of relatively greater significance in the U.S.A. than in countries less well endowed with natural gas. In other countries triethanolamine is the most important of the ethanolamines. This was the case in the U.S.A. until about 1950.

A traditional ethylene oxide application, which is no longer economic, was in the production of acrylonitrile and acrylates. This involved the use of ethylene cyanohydrin as intermediate. The reactions were:

$$\underset{\text{(ethylene oxide ring, }CH_2-CH_2\text{ bridged by O)}}{CH_2-CH_2} + HCN \rightarrow \underset{\text{ethylene cyanohydrin}}{HOCH_2CH_2CN}$$

$$\underset{\text{ethylene cyanohydrin}}{HOCH_2CH_2CN} \rightarrow \underset{\text{acrylonitrile}}{CH_2{=}CHCN} + H_2O$$

$$\underset{\text{ethylene cyanohydrin}}{HOCH_2CH_2CN} + CH_3OH + H_2SO_4 \rightarrow \underset{\text{methyl acrylate}}{CH_2{=}CHCOOCH_3} + NH_4HSO_4$$

$$\underset{\text{ethylene cyanohydrin}}{HOCH_2CH_2CN} + H_2O + H_2SO_4 \rightarrow \underset{\text{acrylic acid}}{CH_2{=}CHCOOH} + NH_4HSO_4$$

Polyethylene

Polyethylene, in its various forms, is one of the major achievements in the field of petroleum chemicals.

The original polyethylene, 'high pressure' or 'low density' polyethylene, came from an accidental discovery in 1933. In two respects the exploitation of the discovery required important technological developments. The ethylene used for high pressure polyethylene required to be purified to an extent not before attempted. In addition, the operation of an industrial process at the pressure required for this reaction posed many design problems. The ethylene is purified to about 99.9 per cent volume, and mixed with 600 parts per million of oxygen (which may be in the form of an organic peroxide). The mixture is compressed to 1 500–2 500 atmospheres. The temperature in the reactor may vary according to requirements from 100–300 °C, but in any particular operation is kept constant. The unreacted ethylene is separated from polyethylene in a separator in which the polymer is in the form of a viscous liquid. This liquid is chilled solid, and is run through a chopper to yield polyethylene flakes. This operation is known as mass polymerization.

Alternatively, solvent polymerization may be used. In this case the reaction is typically carried out at 1 000 atmospheres in the presence of an aromatic hydrocarbon. Purified ethylene with 20 parts per million of oxygen is dissolved in a benzene–water mixture at 1 000 atmospheres and 200 °C. The mixture passes to the reactor, and the oxygen catalyst concentration is maintained by adding water containing 100 parts per million of oxygen. The reactor effluent passes to a separator, and the liquid phase is subsequently distilled to remove the solvent from the polymer.

Variations in pressure, temperature and other conditions, and a variety of subsequent processing steps, will modify the molecular weight, and hence the properties, of the finished product. The polyethylene from this type of process was originally called high pressure polyethylene, but in view of the proliferation of processes using a variety of operating conditions, it is now normally referred to as low density polyethylene. The typical density of this product is 0.92 g/cm^3 The 'weight average'[1] molecular weight of the commercial products may vary from 50 000 to 300 000.

[1] 'Weight average' molecular weight is calculated on the weight proportion of material of different molecular weights present.

The development of the low pressure process, based on Ziegler catalysts, during the 1950's widened the range of products available. The average density of such products is 0.96 g/cm^3 and this is described as high density polyethylene. 'Weight average'[1] molecular weight may vary from 50 000 to 3 000 000 though the very high figures are rarely encountered in practice. High density polyethylene products have appreciably different characteristics from low density polyethylene, including a greater rigidity and a higher softening temperature. The high and low density types of polyethylene are, to a significant extent, complementary rather than competitive.

A Ziegler catalyst system is complex and variable. For the production of low pressure polyethylene, a typical combination is triethylaluminium reacted with titanium tetrachloride.

The catalyst contacts the ethylene stream in an unreactive hydrocarbon medium, such as n-hexane or n-heptane. The titanium trichloride, which is one of the products of the complex catalyst reaction, is insoluble in the hydrocarbon, and is present as a dispersed solid phase. The purified ethylene has appreciable solubility in the hydrocarbon. Typical polymerization conditions are 6–7 atmospheres pressure and 100–170 °C.

The separation and purification of the polymer presents a problem in this case, because any traces of catalyst left in the product will render it unacceptable. It is possible to use a number of different purification treatments. The solvent may be flashed off the polymer, which is then treated with water to kill the catalyst. Alternatively, the catalyst may be deactivated with alcohol. The polymer is then filtered from the solvent mixture and dried. The dry polymer is extruded into strands, cooled and cut into pellets.

A third process for polyethylene manufacture was also developed to the commercial stage in the 1950's. This is the Phillips process. In the standard operation, pure ethylene passes to a reactor with an inert hydrocarbon such as pure dry cyclohexane. Here the mixture contacts the catalyst at a pressure of 20–30 atm., and a temperature of 125–175 °C. The catalyst is of the metal oxide type, typically hexavalent chromic oxide on a silica-alumina support. The reaction products pass to a separator, where ethylene is drawn off and recycled. The polymer is dissolved in the cyclohexane so that the hot solution may be centrifuged or filtered to remove the catalyst. The polymer is precipitated, mechanically separated and dried. A modification of this process employs a liquid medium in which the polymer does not dissolve, but exists as a dispersion. In this case it is necessary for the production of polymer per unit of catalyst to be very high, so that the product is effectively ash-free. This 'particle form' technique offers a more flexible and possibly more economical basis for this process. The Phillips process accounts for nearly 60 per cent of U.S. capacity and 25 per cent of west European capacity for high density polyethylene.

A further process of this type is operated in the U.S.A. Here again the liquid medium in which the reaction takes place will dissolve the polymer as it is formed, enabling it to be easily separated from the catalyst which is of the metal oxide type. The most notable difference from the Phillips process appears to be the higher pressure that is used in this case, 35–68 atmospheres.

In recent years process variants of the Ziegler technique have been numerous. Companies including Hoechst, Solvay, Stamicarbon (DSM), Montedison and Mitsui have announced processes in which the main factor appears to be a modification of the catalyst. In particular, there are two aims of these developments—to decrease the catalyst requirement per unit of polymer produced, to a point at which it is no longer necessary to remove the catalyst from the final product, and by greater control of molecular weight distribution, to give a wider effective range of polymer products, and to minimize the low molecular weight wax by-product. Indeed, for one of the Stamicarbon processes it is claimed that there is no wax produced at all.

The effect of all this process development is to ensure that there is available a wide range of polyethylenes with a spread of such properties as density, softening point and rigidity. The characteristics of the individual grades are commonly established by a property known as melt index, which is related inversely to molecular weight. The melt index represents the number of grams of polymer which will flow through a standard orifice under a standard applied force in 10 minutes at 190 °C.

Apart from the range of grades as produced, modified properties can also be achieved by blending or certain finishing operations. Furthermore, some chemical modifications exist, notably from the chlorination or chlorsulphonation of polyethylene.

The growing place of polyethylene in the world plastics scene may be illustrated by the following figures:

	1955	*1960*	*1970*	*1975* (est.)	*1980* (est.)
World consumption of polyethylene (in '000 tons)	222	956	6 600	13 000	23 100
Share of total plastics consumption %	1	14	22	25	26

Polyethylene is also the greatest single consumer of ethylene, accounting for about 36 per cent of U.S. and almost 50 per cent of west European ethylene consumption.

In the United Kingdom low density polyethylene was produced to the extent of 306 000 tons in 1970 and home consumption was 262 000 tons. The breakdown of United Kingdom usage has changed during the 1960's as follows:

	1960	*1970*
Film and sheet	32%	55%
Injection moulding	25%	15%
Wire and cable	15%	9%
Blow moulding	12%	10%
Pipe and conduit	6%	2%
Extrusion coating	6%	5%
Miscellaneous	4%	4%

The production in the U.S.A. in 1970 was about 1.7 million long tons and home consumption 1.61 million long tons. Almost 50 per cent of the consumption was for film and sheet. If one adds the related extrusion coating market, the combined share of the U.S. low density polyethylene market is over 60 per cent.

Turning to high density polyethylene, the United Kingdom production was marginally over 60 000 long tons in 1970 and the home consumption was 55 000 long tons. The distribution of this amongst different types of application is given in the table below:

	1966	*1970*
Blow moulding	48%	50%
Injection moulding	29.5%	28%
Pipe and profiles	2.5%	7%
Film and sheet	5%	6%
Monofilament	9%	5%
Wire and cable	3%	1%
Miscellaneous	3%	3%

In the U.S.A. the 1970 production figure for high density polyethylene was about 760 000 long tons and the home consumption 723 000 tons. Of this some 45 per cent was for blow moulding and 22 per cent for injection moulding. The pattern of consumption in west Germany is rather different, being quoted for 1969 as including only 35 per cent for blow moulding and 50 per cent for injection moulding. An important factor here was a tremendous short term demand for injection mouldings for packaging use, notably in the form of bottle crates.

Apart from the significance of polyethylene from a manufacturing standpoint, the development of this range of products wrought a revolution in the fields of packaging and flexible containers. A number of the applications for polyethylene derive from the outstanding electrical properties of these materials. The high density grades, apart from their outstanding development in blow moulded bottles, enter the field of semi-rigid mouldings, where they offer competition on the one hand to polypropylene, and on the other to long established products in the polystyrene range.

A new outlet which may become of major importance is the production of a synthetic pulp for paper making. This may be applied to polyethylene or polypropylene. While previous attempts to reproduce paper-like qualities have tended to be based on techniques for modifying plastic films, the current process, which stems from Japan, proceeds directly from the monomer to a fibrous pulp of polyethylene. This pulp is subsequently treated like wood pulp in the paper-making process. The particular merit claimed for this operation is that a plastic-based paper can be produced to compete with standard grades of paper, such as newsprint. Up till now 'plastic paper' has always been a speciality product.

An additional range of potentialities can be envisaged in the various modifications which may be made to the polymer molecules. An example is to be found in the ionomer resins of du Pont. Carboxyl groups are introduced which attach themselves to the main hydrocarbon chains, and metallic cations which interlink the polymer chains. Amongst many changes in polymer properties achieved by this means, perhaps the most notable is transparency.

Styrene

The first step in commercial styrene production is the manufacture of ethyl benzene.

$$\underset{\text{benzene}}{C_6H_6} + \underset{\text{ethylene}}{C_2H_4} \rightarrow \underset{\text{ethyl benzene}}{C_6H_5CH_2CH_3}$$

This is commonly achieved by the alkylation of dry 99 per cent benzene, using an ethylene stream of 95 per cent purity or more in contact with a catalyst of aluminium chloride or phosphoric acid. The most common operation is carried out in the liquid phase, as a Friedel-Crafts reaction using aluminium chloride. Reaction conditions are 85–95 °C and a slightly elevated pressure. The ratio of ethylene to benzene (0.6 mol ethylene to 1 mol benzene) is chosen to give the maximum yield of monoethyl benzene, and to minimize production of higher alkyl benzenes. A small quantity of ethyl chloride is associated with the ethylene feed. The ethyl chloride breaks down into ethylene and hydrogen chloride, and the latter acts as a promoter to the aluminium chloride catalyst.

Once the catalyst complex has been removed from the products of the alkylation reaction in a separator, the remaining hydrocarbon layer is washed with sodium hydroxide solution and water. The remaining mixture contains unreacted benzene, monoethyl benzene and a complex mixture of higher alkyl benzenes and tar. This is separated in a series of distillation columns. Benzene is taken overhead from the first column, and recycled to the alkylator. Monoethyl benzene is the overhead product of the second column, operated at 200 mmHg. The third column operates at 40 mmHg. The bottom product of the third column is the crude 'tar'. The overhead product is a mixture of polyethyl benzenes, which may either be treated in a high temperature dealkylator at about 200 °C (producing benzene and ethyl benzene which are returned to the system) or may be returned to the alkylator, where it is dealkylated, and suppresses the formation of fresh polyethyl benzenes, by virtue of the equilibrium of the reaction (this latter technique may be applied with the aluminium chloride catalyst, but not with phosphoric acid).

Where ethylene and benzene are reacted in the vapour phase over a phosphoric acid catalyst (or alternatively, silica-alumina) the operating conditions used will be 300 °C and 40 atm. pressure, with a ratio of ethylene to benzene of 0.2 : 1.

In these processes a concentrated ethylene stream is generally assumed, but a variant of this type of operation, the Alkar process, is designed to produce ethyl benzene from gas streams containing as little as 10 per cent of ethylene. This is operated commercially on a modest scale.

As a means of obtaining ethyl benzene one must not ignore the superfractionation techniques by which it is recovered from C_8 aromatic concentrates arising from catalytic reforming. These have made a significant impact on ethyl benzene production, notably in the U.S.A. Reference to the processing involved is made in Chapter 12.

The ethyl benzene then requires to be dehydrogenated to styrene:

$$\underset{\text{ethyl benzene}}{C_6H_5CH_2CH_3} \rightarrow \underset{\text{styrene}}{C_6H_5CH{=}CH_2} + H_2$$

This is a strongly endothermic reaction, accompanied by an increase in volume. Optimum conditions will, therefore, include a high temperature and low pressure. The reaction temperature, designed to achieve a satisfactory reaction rate, but to avoid excess degradation, is between 600 and 660 °C. The reduction in pressure is achieved in practice by the use of superheated steam (2.6 lb/lb ethyl benzene) as a diluent. The catalyst used is metallic oxide in character. One alternative is essentially a mixture of magnesium and iron oxides, and others are based on zinc oxide or on iron oxide promoted with chromic and potassium oxides. These are in the form of hard pellets. The reaction products are first cooled against incoming ethyl benzene, and then by steam in heat exchangers. A spray cooler lowers the product temperature to 105 °C, and condenses out tars. The fully condensed reaction product typically contains styrene 35–37 per cent weight, ethyl benzene 59–61 per cent, toluene 1–2 per cent, benzene 0.5–2 per cent and tars 0.2–0.5 per cent.

The gas stream, consisting of impure hydrogen, may be used as fuel or for other chemical reactions.

The separation problems arising from the composition of the reaction product are complex, and involve distillation at fairly high vacuum (i.e. a very much reduced pressure). It is necessary, as a preliminary, to dissolve a small proportion of sulphur in the crude styrene to act as a polymerization inhibitor. In the first column, or dual-column system, benzene and toluene are taken overhead and separated. The benzene is returned to the ethyl benzene alkylator. The separation of styrene and ethyl benzene involves a two- or three-column system operated under vacuum. Ethyl benzene is taken overhead from the first or second column, according to the system used. The final styrene column provides pure styrene as the top product, and a tarry residue at the bottom. The residue may be burned, or passed to a styrene recovery system.

Styrene commonly incorporates a trace of p-tert. butyl catechol to act as a polymerization inhibitor.

It is to be expected that a basic process of this kind will have been subjected to enormous study with the aim of improving economics. The conventional limitation of conversion of ethyl benzene to styrene at around 35–40 per cent, is aimed at the avoidance of degradation of ethyl benzene to benzene and toluene. Modified catalysts are now claimed to offer conversions of up to 55 per cent in this operation. Requirements of steam and other utilities may be minimized by this means, and by close attention to the energy balance of the whole styrene recovery system.

Apart from the conventional dehydrogenation of ethyl benzene to styrene, there are two alternative operations involving ethyl benzene oxidation, which deserve a mention. An obsolete commercial operation started with the oxidation of ethyl benzene to a mixture of acetophenone and phenyl ethyl alcohol. The acetophenone is hydrogenated to give further phenyl ethyl alcohol. This alcohol will dehydrate readily to styrene.

Another unconventional route which has been proposed but not commercially operated, is an oxidative dehydrogenation using sulphur dioxide. This could potentially offer a much higher conversion per pass than current methods. Sulphur is normally present in crude styrene as a polymerization inhibitor in any case, and is consequently an acceptable impurity.

The most successful competitor to the chlorohydrin route to propylene oxide combines the oxidation of propylene with an indirect dehydrogenation. One manifestation of this can dehydrogenate ethyl benzene to styrene. This process is the Halcon process, which is essentially aimed at propylene oxide production, and it is described under this heading in the next chapter.

Styrene is one of the most important products in the petroleum chemical field. Production in the U.S.A. in 1970 was nearly 2 million long tons and by 1975 the total will be over 3 million long tons.

Applications for styrene in the U.S.A. in 1970 were said to be:

General purpose polystyrene	25%
High impact (rubber modified) polystyrene	20%
Styrene-butadiene elastomer	17%
Exports	16%
ABS and SAN resins	9%
Styrene-butadiene copolymer resins	6%
Styrenated polyesters	4%
Miscellaneous	3%

The styrene polymers (including ABS and SAN resins) account for two-thirds of western European styrene consumption, in 1969 almost 900 000 tons of styrene being consumed in these polymers.

The Japanese styrene demand has been presented as follows (in '000 long tons):

	1970	*1974*
For polystyrene (inc. foams)	478	826
SB rubber	110	163
Polyester resins	41	72
ABS resin	90	176
Miscellaneous	13	23
Exports	62	26
Styrene demand	794	1 286

Polystyrene production in the United Kingdom in 1970 was 151 000 long tons, and home consumption 140 000 tons. Specific fields of application for this material were:

Packaging	51%
Household and premium goods	10%
Toys	10%
Appliances	8%
Sheet (excluding packaging)	5%
Bathroom accessories	3%
Shoe heels	3%
Light fittings	2%
Combs and brushes	1%
Miscellaneous	7%

To this total should be added the production in the United Kingdom of 25 000 tons of expanded polystyrene in 1970, and a home consumption of nearly 16 000 tons. This tonnage was more or less equally split between building insulation, retail sales (e.g. ceiling tiles) and packaging applications. Whilst the tonnage of these products does not look impressive, the product is so light that the volume of this production is enormous.

The U.S.A. produced over 1 million tons of polystyrene resin in 1970, of which a high proportion found application in moulding and extrusion. Expandable polystyrene amounted to almost 100 000 long tons.

Detailed figures are given for Japan as follows (in '000 tons):

	1970	*1971* (est.)	*1972* (est.)
Industrial demand			
T.V.	27.6	30.6	33.0
Refrigerators	16.5	18.8	21.2
Radio	10.8	11.7	12.5
Electrical appliances	51.1	63.0	75.7
Machinery	20.0	23.9	27.6
Consumer goods			
Containers, kitchenware	145.0	174.5	207.4
Toys	22.8	25.3	27.8
Writing materials	17.0	24.0	31.0
Exports	60.0	63.0	66.0
Total polystyrene demand	370.8	434.8	502.2

Polystyrene can be made in a number of different ways. Continuous mass polymerization gives a product of excellent clarity but less satisfactory moulding properties. Solution or suspension polymerization gives much better control of molecular weight, and hence improved moulding qualities, but trace impurities tend to impair the clarity.

Polystyrene itself tends to be brittle, and much attention has been paid to the production of 'high impact' grades. This is normally effected by incorporating 10–12 per cent of natural or synthetic rubber (frequently polybutadiene) into polystyrene.

Modified polymers are produced by the copolymerization of styrene with such monomers as acrylonitrile or butadiene. Such copolymers are designed to combine the desirable moulding properties and cheapness of polystyrene with specific requirements for a variety of uses. This is one of the faster moving sectors of the styrene market. The most important of the copolymers is known as ABS resin, since the co-monomers are acrylonitrile, butadiene and styrene. In 1970 some 250 000 long tons of ABS resins, and the related but less important SAN resins, were made in the U.S.A. It has been suggested that U.S. demand by 1975 could reach 500 000 long tons. In the United Kingdom in 1970 production of ABS resins was 28 000 long tons and consumption 20 000 tons. This resin finds wide application in all forms of appliances.

The composition of the styrene copolymer resins can vary but ABS resins may typically be expressed in weight proportions as 50 per cent styrene, 30 per cent acrylonitrile and 20 per cent butadiene. The SAN resins typically consist of 70 per cent weight styrene and 30 per cent acrylonitrile.

By the incorporation of about 10 per cent of a volatile hydrocarbon, such as pentane or butane, into polystyrene granules or beads, the beads can be expanded into a low density block or board. This material has exceptionally good thermal insulating properties. The beads can be treated with a flame-retarding compound, which will eliminate the hazard of flammability of the finished board. This is obviously important where the insulating board is to be used in housing construction.

A technique of some note in the production of foamed polystyrene sheet makes the material direct from general purpose crystal polystyrene. This requires a preliminary dry compounding of the crystal with a nucleating agent which will not be destroyed by the heat of extrusion (sodium bicarbonate or talcum powder will do). The pellets are melted to a homogeneous mass in the normal extrusion equipment, and at a certain point an expanding gas (e.g. pentane) is introduced directly into the extruder barrel. The gas concentrates round the dispersed particles of the nucleating agent. The release of pressure as the material leaves the extruder allows the gas to expand and create a uniform foam.

The other major application of styrene in the production of the standard (SBR) type of synthetic rubber is considered in more detail later. This application represents a declining share of the styrene market since the advent of new synthetic rubbers is stunting the growth of SBR.

Ethyl Chloride

Ethylene is the major source of ethyl chloride, although some is produced from both ethane and ethyl alcohol. The reaction is a straightforward hydrochlorination.

$$CH_2{=}CH_2 + HCl \rightarrow \underset{\text{ethyl chloride}}{CH_3CH_2Cl}$$

The reaction may proceed in the vapour phase, usually at temperatures of 130–250 °C, in the presence of aluminium chloride as catalyst. Alternatively, at 35–40 °C, and pressures sufficiently above atmospheric to maintain the reaction products in the liquid phase, using ethylene dichloride (possibly with some ethyl chloride) as a solvent, the reaction will proceed with aluminium chloride as catalyst. The heated products are fed into a column or flash drum, where ethyl chloride is separated from the less volatile polymer by-products. The ethyl chloride is then purified by fractionation.

As was indicated in Chapter 6, 90 per cent of the ethyl chloride made in the U.S.A. in 1970 was used in the production of tetraethyl lead, which is added to gasoline to improve its anti-knock characteristics. The important developments in lead additives for gasoline are briefly discussed in Chapter 6.

Ethylene Dichloride

The reaction between chlorine and ethylene is an exothermic addition reaction

$$CH_2{=}CH_2 + Cl_2 \rightarrow \underset{\text{ethylene dichloride}}{CH_2ClCH_2Cl}$$

which will take place either in the vapour or liquid phase.

Commercial operation in the vapour phase requires the presence of a catalyst, which takes the form of a metallic chloride, and may conveniently be iron chloride, which is likely to be present in any case. The chlorination operates by a free-radical chain mechanism, and the catalyst plays a part in releasing free chlorine radicals. The temperature needs careful control, and should not rise much above 125 °C or the selectivity of the reaction will be impaired.

It is more common to carry out the chlorination in the liquid phase. It is convenient to use ethylene dichloride itself as a solvent in the reactor, and the reaction takes place at about 50 °C and approximately atmospheric pressure. It is commonly indicated that a ferric chloride catalyst should be added, though in fact ferric chloride will inevitably be produced *in situ* in quantities which are normally adequate. Temperature control is important, or excessive quantities of more highly chlorinated products may be formed. The liquid products from the reactor are washed with dilute sodium hydroxide solution to remove hydrochloric acid. Ethylene dichloride is purified by distillation.

While there are advantages in using concentrated ethylene streams for these operations, it is possible to operate both the vapour and liquid phase processes with dilute ethylene streams. Two processes for vinyl chloride production from naphtha which illustrate this are the S.B.A. and the Kureha process. In each case naphtha is cracked to provide an equimolecular mixture of ethylene and acetylene. The acetylene in the stream is reacted with hydrogen chloride to produce vinyl chloride. In the S.B.A. process the vinyl chloride is separated from ethylene by an independent solvent extraction system. The dilute ethylene stream is then reacted with chlorine in the vapour phase. In the Kureha process vinyl chloride is absorbed in ethylene dichloride produced downstream in the system. The ethylene stream, which contains hydrogen, methane and oxides of carbon, is reacted with chlorine at 50–70 °C and 4–5 atm. In each of these processes it is desirable that ethylene should be present in slight excess, to minimize the amount of unreacted chlorine in the effluent from the reactor.

A minimal quantity of ethylene dichloride still occurs as a by-product of the chlorohydrin process to ethylene oxide.

In the past few years the whole outlook for ethylene dichloride, and for the ethylene route to vinyl chloride, has been revolutionized by the development of the oxychlorination routes. The oxychlorination process uses hydrogen chloride and oxygen (or air) as the chlorination medium to produce ethylene dichloride from ethylene. The effect of this is to put hydrogen chloride economically almost on a par with chlorine. Hydrogen chloride occurs as a by-product of many operations, notably in the pyrolysis of ethylene dichloride to vinyl chloride. Previously, much of this material could not be effectively used since there was no satisfactory way of converting the hydrogen chloride back to chlorine. It is this problem that has been obviated by the development of oxychlorination techniques.

The oxychlorination catalyst used is cupric chloride modified by the addition of potassium chloride and supported on an inert material such as alumina or silica. The presence of the alkali metal chloride reduces the tendency of the catalyst to volatilize at the reaction temperature used. Since the oxychlorination is a highly exothermic reaction, special precautions must be made to maintain control of temperature.

The temperature of the reaction is around 250–315 °C and the pressure may be atmospheric or rather higher. The feed streams should be held approximately in the stoichiometric ratio of 1 ethylene: 2 hydrogen chloride: 0.5 oxygen.

$$CH_2{=}CH_2 + 2HCl + \tfrac{1}{2}O_2 \rightarrow CH_2ClCH_2Cl + H_2O$$

There are many oxychlorination processes now in use and most of them operate in the vapour phase. The Kellogg process, however, operates in the liquid phase. This proposes a temperature of 170–185 °C and a pressure of 15–20 atm. The reaction medium is an aqueous solution of the cupric chloride used as catalyst. It is claimed that with this reactor system it is possible to carry out both oxychlorination and chlorination of ethylene in the same reactor.

Vinyl chloride is now well established as the predominant outlet for ethylene dichloride. In the U.S.A. in 1970 some 75 per cent of ethylene dichloride consumption was for vinyl chloride; 9 per cent was converted to chlorinated solvents, 4 per cent to ethylene amines, a mere 3 per cent was used as a lead scavenger in tetraethyl lead fluid, and the remaining 9 per cent found miscellaneous solvent uses or went for export.

Published production figures for ethylene dichloride are sometimes low, since the product is not always isolated. Te U.S. figure for 1970 was around 2.8 million long tons. The demand for ethylene dichloride in the U.S.A. has been forecast to rise to 5 million tons by 1975. This assumes the virtual elimination of acetylene-based vinyl chloride, and vigorous growth for both vinyl chloride polymers and for the ethylene-based chlorinated solvents.

Figures for vinyl chloride and PVC production have been quoted in Chapter 4.

As has already been mentioned (in Chapter 4) the traditional process for making vinyl chloride from acetylene has largely been superseded. Such acetylene-based vinyl chloride as still exists is mostly produced by the 'balanced process'. In this, hydrogen chloride, which is produced in the pyrolysis of ethylene dichloride to vinyl chloride, is added to acetylene to make more vinyl chloride. In short, acetylene can by this means be used as a scavenger for hydrogen chloride. This process was an intermediate step, before the oxychlorination processes became widely used, whereby some advantage could be taken of the cheapness of ethylene as a raw material, without any net production of unwanted hydrogen chloride. Today such hydrogen chloride is fed to the oxychlorination process.

The production of vinyl chloride from ethylene dichloride involves the removal of a hydrogen chloride molecule.

$$\underset{\text{ethylene dichloride}}{CH_2ClCH_2Cl} \rightarrow \underset{\text{vinyl chloride}}{CH_2{=}CHCl} + HCl$$

This dehydrochlorination may be achieved either by pyrolysis, or by treatment with sodium hydroxide.

The pyrolysis process passes ethylene dichloride at 480–500 °C and 3 atmospheres pressure over a catalyst of pumice, kaolin or carbon, though a catalyst is not universally used. The exit gases are quenched in a stream of condensed unreacted ethylene dichloride. Uncondensable gases are scrubbed with water to recover hydrogen chloride. Vinyl chloride is recovered from the liquid mixture by distillation under 5 atmospheres pressure, and is purified by redistillation.

The alternative process requires ethylene dichloride to be heated in the presence of sodium hydroxide.

$$CH_2ClCH_2Cl + NaOH \rightarrow CH_2{=}CHCl + NaCl + H_2O$$

The reaction takes place at 10 atmospheres pressure and 140–150 °C. Products overflowing from the reactor are cooled and pumped to a distillation column operating under pressure. From the column bottoms, unreacted ethylene dichloride is recovered and recycled. Alkaline decomposition of ethylene dichloride is increasingly out of favour today, since chlorine is effectively lost as sodium chloride.

A consideration of the two steps, oxychlorination of ethylene to ethylene dichloride, and pyrolysis of the ethylene dichloride to vinyl chloride, clearly indicates the possibility of condensing these two operations into a one-step process. Several companies have taken out patents on the subject. I.C.I., for example, envisage ethylene, hydrogen chloride and oxygen being reacted over a modified Deacon catalyst (cupric chloride/potassium chloride supported on diatomaceous earth) at nearly 500 °C to give a fairly high yield of vinyl chloride. Such processes have obvious attractions, but have not yet become fully commercialized.

The aim to produce vinyl chloride directly from ethylene and chlorine has recently been directed more to reactions taking place in the absence of oxygen. A process of Pechiney-St. Gobain, which has been taken to the commercial stage, starts with a high temperature chlorination of ethylene in the absence of a catalyst. Conditions in the adiabatic reactor are 370–500 °C and 1.5 atm. The products of this reaction are vinyl chloride and dichloroethylene, together with hydrogen chloride.

Dichloroethylene is fed to a cold chlorination unit. The reaction is non-catalytic and takes place at 50 °C and atmospheric pressure. The product is tetrachloroethane, which is thermally cracked to trichloroethylene and hydrogen chloride.

Hydrogen chloride and unconverted ethylene from these two process units, together with fresh hydrogen chloride and ethylene, are then fed, together with air, to an oxychlorination unit. Here the reaction is promoted by a fluidized catalyst comprising an active copper salt on an oxide substrate. The reaction temperature is 280–480 °C and pressure 2–8 atm. The products are water and a mixture of di-, tri-, and tetrachloroethane. In the specific plant operation at Saint Auban, France, only dichloroethane is recovered in a pure state, but it is possible to recover trichloroethane and tetrachloroethane as pure compounds if required.

A small scale unit, producing vinyl chloride and chlorinated solvents in the same plant complex, is operated in Japan by Toagosei Chemical Co. The first step is a liquid phase chlorination in which ethylene passes through chlorine dissolved in a mixture of chloroethanes. No catalyst is employed, the temperature is 100–130 °C and the pressure about 8 atm. The reaction medium is circulated through an external cooler to remove heat of reaction. The product is composed of a mixture of chloroethanes, together with hydrogen chloride. Di- and trichloroethanes are recycled to the reactor.

The higher chloroethanes are cracked at 430–450 °C and 9 atm. There is a high degree of selectivity in the decomposition, favouring the formation of trichloroethylene and perchloroethylene. The cracker effluent is quenched, the co-product hydrogen chloride passes off, and trichloroethylene and perchloroethylene are purified by distillation. The hydrogen chloride is used to feed an oxychlorination unit producing ethylene dichloride (1,2-dichloroethane) from which vinyl chloride may be produced in the conventional way. An additional variant is that 1,1,2-trichloroethane may be separated from the chloroethanes to produce vinylidene chloride and methyl chloroform (see Chapter 4).

The chlorination of ethylene dichloride (which may be a direct chlorination, an oxychlorination or a combination of the two), followed by raising the reaction products to pyrolysis temperature, can form a group of chlorinated hydrocarbon solvents, of which the major ones are perchloroethylene, carbon tetrachloride and trichloroethylene. The temperatures employed in these reactions are 360–450 °C according to the circumstances and the product required. The oxychlorination processes require a modified cupric chloride catalyst. Processes of this kind have assumed commercial importance in the U.S.A.

$$CH_2ClCH_2Cl + 2Cl_2 \rightarrow CCl_2{=}CHCl + 3HCl$$
$$CH_2ClCH_2Cl + 3Cl_2 \rightarrow CCl_2{=}CCl_2 + 4HCl$$
$$CH_2ClCH_2Cl + 5Cl_2 \rightarrow 2CCl_4 + 4HCl$$

The above equations represent the overall effect of these processes in simplified form. The actual chemistry involves two steps, which may take place within the same reactor. Firstly ethylene dichloride with chlorine (or its oxychlorination equivalent) is converted to the more highly chlorinated ethanes. These are then cracked to produce the unsaturated derivatives and hydrogen chloride. From tetrachloroethane is derived trichloroethylene, and from pentachloroethane comes perchloroethylene. This part of the chemical operation is essentially the same as in the derivation of these products from acetylene.

The production of chlorinated solvents from ethylene has been developed extensively during the past few years, and this is becoming another field in which acetylene, as a raw material, is on the defensive. The main developments relate to perchloroethylene and trichloroethylene.

A recent trend has been towards the use of oxychlorination in this field. Here, as in vinyl chloride manufacture, oxychlorination acts as a balancing factor, ensuring that the hydrogen chloride co-product can always find an outlet, at a value virtually equivalent to its chlorine content.

The number of process variations now available in this field makes it possible to vary the ratios of the chlorinated solvents produced over a wide range. A further merit of some of these processes is that chlorinated by-product streams (with products up to two carbon atoms at least) can be fed into the reaction, along with ethylene dichloride.

Ethylene dichloride may alternatively be chlorinated to 1,1,2-trichloroethane. This may be conventionally dehydrochlorinated to vinylidene chloride and hydrochlorinated again to produce 1,1,1-trichloroethane (methyl chloroform).

A product which has assumed some significance in this field is ethylidene chloride. This is 1,1-dichloroethane, in contrast to ethylene dichloride which is 1,2-dichloroethane. Ethylidene chloride can occur in various chloroethane streams, or can be made by the hydrochlorination of vinyl chloride. The chlorination of ethylidene chloride at 315–420 °C produces 1,1,1-trichloroethane (methyl chloroform) directly. This whole field is one in which hydrogen chloride is frequently involved as a product or a reactant, and consequently producers of 1,1,1-trichloroethane are commonly also involved in oxychlorination processes of one kind or another.

A further reaction of some commercial importance is that between ethylene dichloride and excess aqueous ammonia under elevated pressure at about 120 °C. The major product is the chemical intermediate ethylene diamine, together with a proportion of higher polyamines.

$$CH_2ClCH_2Cl + 4NH_3 \longrightarrow CH_2NH_2CH_2NH_2 + 2NH_4Cl$$

This group of ethylene amines has a wide range of applications, including chemical intermediates in the production of sequestering agents, speciality polymers and pesticides.

A rather similar operation, in which ethylene dichloride is reacted with rather less ammonia, gives rise to ethylene imine.

$$CH_2ClCH_2Cl + NH_3 \longrightarrow \underset{\diagdown\;NH\;\diagup}{CH_2{-}CH_2} + 2HCl$$

This product has been widely canvassed as a versatile chemical intermediate. It remains a fairly specialized product, with a number of useful derivatives employed in paper processing. Ethylene imine is also made by the dehydration of monoethanolamine.

Mention should also be made of the polysulphide rubbers. These are made by reacting ethylene dichloride with sodium polysulphide.

They have excellent resistance to oil and solvents, but poor heat resistance and an unpleasant odour. Their specialized applications are on a somewhat limited scale—in caulking and sealing, moulded gasketing and sealing components, and hose linings.

While it may be claimed that ethylene dichloride is a versatile chemical intermediate, the vinyl chloride application is dominant and likely to remain so. Polyvinyl chloride is made on a vast scale and, although its growth rate will probably not maintain the annual figure of 17 per cent achieved in the Common Market countries during the 1960's, a future growth rate of 10 per cent a year seems likely. Recent forecasts of consumption in 1975 included the following estimates:

Common Market countries and United Kingdom	3.79 million long tons
U.S.A.	2.12 million long tons
Japan	1.63 million long tons

The per capita consumption is exceptionally high in west Germany, the Netherlands and Japan.

Ethylene Dibromide

This process is analogous to that producing ethylene dichloride, and is the only major industrial use for bromine (commonly recovered from sea water).

The addition of bromine to ethylene gives ethylene dibromide.

$$CH_2{=}CH_2 + Br_2 \rightarrow CH_2BrCH_2Br$$

The process conditions are similar in principle to those for ethylene dichloride.

Ethylene dibromide is an effective lead scavenger in tetraethyl lead fluid as a gasoline additive. It is used in automotive gasoline in conjunction with ethylene dichloride, which is, of course, much cheaper. At one time ethylene dibromide was used in the aviation gasoline additives, to the exclusion of ethylene dichloride, but the advent of jet fuels largely annihilated this market. With the present concern over the pollution aspects of lead-containing additives to gasoline (discussed in relation to ethyl chloride) the future position of ethylene dibromide must be open to some doubt. There are minor alternative agricultural applications in nematode control (described as 'soil fumigation') and grain fumigation.

Acetaldehyde (Wacker process)

The Wacker process must now be ranked among the major processes for acetaldehyde manufacture.

The oxidation of ethylene to acetaldehyde involves the use of palladium chloride as catalyst. The basic reactions involving palladium chloride are:

$$CH_2{=}CH_2 + PdCl_2 + H_2O \rightarrow \underset{\text{acetaldehyde}}{CH_3CHO} + Pd + 2HCl$$

$$Pd + 2HCl + \tfrac{1}{2}O_2 \rightarrow PdCl_2 + H_2O$$

The first reaction is rapid and the second relatively slow. Cuprous chloride is therefore introduced, in effect, as a promoter for the second of the above reactions

$$2CuCl_2 + Pd \rightarrow 2CuCl + PdCl_2$$

$$2CuCl + 2HCl + \tfrac{1}{2}O_2 \rightarrow 2CuCl_2 + H_2O$$

The net reaction and regeneration steps may be expressed as follows, leaving the palladium chloride unchanged:

$$CH_2{=}CH_2 + 2CuCl_2 + H_2O \xrightarrow{PdCl_2} \underset{\text{acetaldehyde}}{CH_3CHO} + 2HCl + 2CuCl$$

$$2CuCl + 2HCl + \tfrac{1}{2}O_2 \longrightarrow 2CuCl_2 + H_2O$$

The process may be carried out as a one-stage operation in the liquid phase, for both reaction and regeneration. This requires a pure ethylene stream for the reaction, and oxygen for the regeneration.

The two-stage operation, also in the liquid phase, can employ any ethylene-rich stream in the reaction stage, and air oxidation in the regeneration stage. In each case the reaction takes place at a temperature of 100 °C or marginally above, and a pressure of around 10 atm.

Alternative processes to acetaldehyde either involve an indirect derivation from ethylene via ethyl alcohol, or the use of acetylene, which is inherently a more expensive raw material than ethylene. There is an obvious economic advantage in a direct high yielding derivation from ethylene. In practice, a 95 per cent yield of acetaldehyde on ethylene has been claimed. This is another illustration of a major guiding principle (the tendency to simplify synthesis processes) in modern industrial chemistry.

One of the major problems encountered in the early operation of this process was the choice of construction materials for the reactor, where the highly acid metal chloride solutions were found to be very corrosive. A solution was found in the fabrication of critical items of equipment in titanium metal.

Earlier expectations of the economics of this process led to the postulation that it could form an alternative route to acrylonitrile. The reactions were:

$$CH_3CHO + HCN \longrightarrow \underset{\text{lactonitrile}}{CH_3CHOHCN}$$

$$\underset{\text{lactonitrile}}{CH_3CHOHCN} \longrightarrow \underset{\text{acrylonitrile}}{CH_2{=}CHCN} + H_2O$$

This has not in fact proved a competitive route to acrylonitrile, although it has been used in the production of lactic acid, which is obtained by hydrolysis of lactonitrile. It has replaced, to a significant extent, the production of lactic acid by fermentation of carbohydrates.

Vinyl Acetate

The proposal that the oxidation of ethylene using the palladium chloride catalyst in the medium of glacial acetic acid would yield vinyl acetate, originated with the Soviet chemist Moiseev. The provision of acetate ions (e.g. the addition of sodium acetate) was found to be necessary. Otherwise the chemistry is similar to that of the Wacker acetaldehyde process.

$$CH_2{=}CH_2 + CH_3COOH + PdCl_2 \longrightarrow \underset{\text{vinyl acetate}}{CH_3COOCH{=}CH_2} + 2HCl + Pd$$

$$Pd + 2CuCl_2 \longrightarrow PdCl_2 + 2CuCl$$

$$2CuCl + 2HCl + \tfrac{1}{2}O_2 \longrightarrow 2CuCl_2 + H_2O$$

The reaction conditions were moderate (up to 100 °C, up to 22 atm.) but the whole operation was bedevilled by corrosion problems arising mostly from hydrogen chloride. To overcome these difficulties by the use of specialized construction materials—titanium, resin-impregnated graphite ceramic linings, for example—is costly and not always completely effective.

One of the original aims of the process was to combine the production of vinyl acetate with the production of acetaldehyde, which is, after all, the original form of ethylene oxidation by the Wacker process. By this means, the whole operation would be independent of external sources of acetic acid and would be entirely ethylene-based. This proved a very sensitive operation to handle, but in the commercial operation of the process acetaldehyde was normally obtained as a by-product.

The liquid phase processes have not proved a success—I.C.I. for example have, surprisingly enough, revamped their unit to make propylene oxide by the chlorohydrin process—and recent developments are concentrating on vapour phase processes. These have been developed almost simultaneously by Bayer in Germany and U.S.I. Chemicals in the U.S.A. There is little obvious difference between the processes.

Ethylene, oxygen and acetic acid are reacted over a fixed catalyst bed in the vapour phase at 175–200 °C and 5–10 atm. pressure. The catalyst consists of a noble metal (usually palladium) distributed in a finely divided state to the extent of 0.1 to 2 per cent by weight on a silica or alumina support. It is advantageous also to add an alkali metal acetate to the catalyst to the extent of 0.5 to 5 per cent by weight. The reaction is highly exothermic. The temperature in the tubes containing the catalyst is controlled by boiling water in the reactor shell. The steam generated is used in the subsequent distillation process.

Reaction products are cooled to 40 °C to condense some 70 per cent of the vinyl acetate. Liquid and gas are separated in two separators (the first at higher pressure, the second at low pressure) and the gases are scrubbed with propylene glycol to recover vinyl acetate present. The gases are then treated to remove carbon dioxide, and the ethylene remaining is recycled to the reactor. Vinyl acetate from the propylene glycol scrubber is separated from propylene glycol by distillation, and is added to the condensate from the gas/liquid separators to proceed to the distillation section. Here vinyl acetate is separated from acetic acid and small quantities of volatile by-products and heavy fractions.

This type of process has far fewer problems with corrosion than the liquid phase process; there are few by-products and the utility demands are moderate. There is now substantial commercial development of vinyl acetate production by this means and every indication that this will become the dominant process in this field.

$$\underset{\text{ethylene}}{CH_2{=}CH_2} + \underset{\text{acetic acid}}{CH_3COOH} + \tfrac{1}{2}O_2 \rightarrow \underset{\text{vinyl acetate}}{CH_3COOCH{=}CH_2} + H_2O$$

Ethylene-Propylene Rubbers (EPR and EPDM)

The concept of producing commercially acceptable rubbers by the copolymerization of ethylene and propylene has obvious attractions. Ethylene and propylene are both economical monomers, and, if a reasonable process can be devised to offer a product of acceptable quality, a highly attractive project should result. A practical point is that the ethylene-propylene rubbers have a lower density than most competitive materials. Since many rubber products (such as tyres) involve a specific volume of rubber, the same dimensions can be produced with a reduced weight, and, presumably, lower raw material cost.

Present developments stem from the same family of Ziegler catalysts as is used for high density polyethylene, and for polypropylene.

Polymerization conditions have been quoted as not more than 60 °C and 25 atmospheres in the DSM process. Alternatively, Montedison indicates temperatures in the range −20 to +20 °C and pressures of 3 to 10 atmospheres. In the latter case the catalyst components include vanadium compounds and aluminium alkyls.

According to the properties required the proportions of ethylene and propylene in the copolymer may be varied. As the whole point of this exercise is to obtain a product different from polypropylene and polyethylene, it is fair to observe that the normal balance of the two monomers will tend to approach an equimolar proportion. As is explained below, the common products today are based on three components, and the average ratio has been given as ethylene 55 per cent by weight, propylene 40 per cent, and the third component (or termonomer) 5 per cent.

The properties of the ethylene-propylene rubbers present some difficulties from the point of view of subsequent processing. In the polymerization, the chain structure takes the place of the double bonds, as exemplified by an ethylene polymer.

$$CH_2{=}CH_2 \xrightarrow{\text{polymerization}} {-}CH_2{-}CH_2{-}CH_2{-}CH_2{-}CH_2{-}$$

Since the molecule is, for all practical purposes, saturated, the standard vulcanization technique, using sulphur, becomes impossible. The less familiar peroxide vulcanization can be applied.

It is commercially desirable to modify the molecule so that normal vulcanization techniques can be employed. Additional double bonds may be introduced into the polymer molecule by copolymerizing with a small proportion of dienes (as is done with butyl rubber). It is necessary to use an unconjugated diene for this purpose. This ensures that individual double bonds can branch out from the main hydrocarbon chain, and form a focus for the vulcanization to proceed. Among the dienes in commercial use in the U.S.A. for this purpose are dicyclopentadiene, 1,4-hexadiene and ethylidene norbornene.

Although the ethylene-propylene 'terpolymers' (designated EPDM) have overcome the problem of vulcanization, it will still be necessary to establish techniques for using the rubber in such specialized processes as tyre manufacture. Ethylene-propylene rubbers are lacking, for instance, in 'tack' (or bonding power), a characteristic desirable for tyre rubbers. Such difficulties can usually be overcome, but always at a cost.

Ethylene-propylene rubbers do not fit neatly into any of the existing categories. They probably have more in common with butyl rubber than any other type, and are more likely to replace general purpose synthetic rubbers (SBR) than natural rubber.

The breakthrough of the ethylene-propylene rubbers into a substantial scale of usage in tyre manufacture is regularly predicted, but has not yet taken place. As a result there has always been a tendency towards overcapacity, especially in the U.S.A. Consumption there was 52 000 long tons in 1970, and it is generally expected that this will at least double by 1975. Buoyant forecasters have even estimated figures as high as 200 000 long tons by 1975. Since current U.S. capacity is in excess of 150 000 long tons a year, the consumption still has a long way to go to catch up. Production and consumption in individual European countries is relatively small. There is currently a significant export from the U.S.A.—their production in 1970 was 75 000 long tons, well in excess of the consumption.

Higher Alcohols and Olefins

This is another aspect of 'Ziegler chemistry'. The basic reaction is between ethylene and aluminium triethyl at high pressure (50–100 atm.) and temperatures below 130 °C.

The products of this reaction are a range of aluminium alkyls, which are oxidized with dried air to form the corresponding alkoxides. The alkoxides are hydrolysed with 98 per cent sulphuric acid to straight chain primary alcohols, suitable for conversion to detergent products which are biodegradable.

$$\underset{\text{aluminium triethyl}}{Al(C_2H_5)_3} + nC_2H_4 \longrightarrow \underset{\text{aluminium alkyls}}{Al\begin{cases} CH_2R^1 \\ CH_2R^2 \\ CH_2R^3 \end{cases}} \xrightarrow{O_2} \underset{\text{aluminium alkoxides}}{Al\begin{cases} OCH_2R^1 \\ OCH_2R^2 \\ OCH_2R^3 \end{cases}}$$

$$Al\begin{cases} OCH_2R^1 \\ OCH_2R^2 \\ OCH_2R^3 \end{cases} \xrightarrow{H_2SO_4} R^1CH_2OH + R^2CH_2OH + R^3CH_2OH + Al_2(SO_4)_3$$

There are plants operating this type of process in the U.S.A. and Germany. The products compete in the alcohol field with fatty alcohols obtained by the reduction of natural fatty acids by sodium or by catalytic hydrogenation, and with primary alcohols obtained by applying the Oxo process (see Chapter 11) to straight chain olefins. There are also secondary alcohols from paraffin oxidation and a whole spate of alternative detergent raw materials. In this technological jungle, the 'Ziegler alcohols' have carved themselves a place by virtue of their almost total lack of branching, and consequently favourable biodegradability and other properties. There is, however, only limited flexibility to adjust the proportions of the various alcohols produced which may typically be from C_6 to C_{26}. The alcohols in the C_8–C_{10} range are used for plasticizers. C_{12}–C_{18} alcohols may be used for detergents. Higher alcohols are used in resins, and as evaporation retarders or lubricants.

As an alternative it is possible to heat the aluminium trialkyls to 300 °C in the presence of ethylene at an elevated pressure. Thermal displacement then occurs, regenerating aluminium triethyl and the side chains split off to give straight chain alpha olefins. These olefins have a special value in such operations as biodegradable detergent manufacture. The alternative straight chain alpha olefin streams (e.g. from wax cracking) are significantly less chemically precise than the 'Ziegler olefins'. All these operations face an economic problem, in that a wide range of olefins of varying chain lengths are produced in one operation, and it is necessary to balance consumption with production on the entire range. The virtues of Ziegler olefins in chemical purity are offset by a rather high cost.

A basically similar process for alpha olefins in the C_4–C_{18} range has been outlined by Mitsui, although it has not yet been commercially developed. In this case the basic polymerization step takes place at 0–20 °C and 10–30 atmospheres. Modifications in the Ziegler catalyst are claimed to be capable of varying the carbon-number distribution of the olefins. One variant concentrates the products in the C_{10}–C_{16} fraction, and another emphasizes the C_4–C_{10} range of olefins.

Acrylic Acid

Acrylic acid would seem to be one of the products already enjoying a wide enough variety of commercial manufacturing processes. Recently an interesting new process development was announced by Union Oil of California. It is a further extension of the Wacker process catalyst system.

Excluding the catalyst complications the basic reaction may be expressed as follows:

$$CH_2{=}CH_2 + CO + \tfrac{1}{2}O_2 \rightarrow \underset{\text{acrylic acid}}{CH_2{=}CHCOOH}$$

The catalyst system is the traditional Wacker process combination of palladium chloride and copper chloride. The reaction is carried out in acetic acid solvent at 140 °C and about 75 atm. This is described as the oxidative carbonylation of ethylene.

There is obvious merit in the possibility of producing acrylic acid in quite good yield from ethylene. There are many problems yet to be overcome before the process can be effectively commercialized. These problems include the formation of by-products and the familiar question of corrosion arising from the Wacker catalyst complex.

Ethylene Alkylate

This is not a chemical product in the normal sense, but it could be a consumer of ethylene. The alkylation of lower olefins with a paraffin—more particularly isobutane—is a means of providing high-octane gasoline. High-octane gasoline components are of particular significance in the current atmosphere of impending change in the composition of gasoline which may become necessary—at least in the U.S.A.—if the conventional lead-based additives have to be eliminated. The olefin alkylates have high 'clear octane' numbers, which is to say they operate as high octane components even in the absence of lead additives. Ethylene alkylate has particular attractions in that it provides a high-octane component at the volatile end of the formulation range.

The process of reacting ethylene and, typically, isobutane, is a fairly standard alkylation operation catalysed by aluminium chloride. The product is not a chemical entity but is composed mostly of C_6 hydrocarbons, notably 2,3-dimethylbutane.

The significance of ethylene alkylate lies in the fact that whilst ethylene is only one of the olefins which may be used as alkylation feedstock, it is the one olefin which is not already available in substantial quantity to the oil refiner. It could happen, therefore, that new ethylene plants could be designed wholly or in part to serve as refinery feedstock. This is a familiar pattern in the U.S.A. with propylene, butylenes and amylenes, but the intrusion of a refinery demand into the ethylene picture would have a profound influence on the supply/demand pattern.

Ethylene alkylate has not yet assumed much commercial significance, since historically ethylene has been a relatively expensive feedstock. It also differs from propylene and the butylenes in that it requires aluminium chloride to catalyse the alkylation, rather than the conventional hydrofluoric acid or sulphuric acid.

Propionic Acid

The process for manufacturing propionic acid from ethylene and carbon monoxide is another example of 'Reppe chemistry'. The reaction is

$$CH_2{=}CH_2 + CO + H_2O \rightarrow CH_3CH_2COOH$$

and this type of reaction is sometimes very loosely described as a 'wet Oxo' process (see Chapter 11).

The ethylene and carbon monoxide are introduced into liquid crude propionic acid which contains 10–15 per cent of water. Nickel carbonyl is used as catalyst, and the reaction pressure is 200–250 atm.

This is an illustration of the way in which the original concept of Reppe chemistry, that concentrated upon high-pressure acetylene reactions, has been adapted to use olefins and methanol as raw materials.

This method is used on a substantial scale in Germany. It is the main competitor to the alternative manufacture of propionic acid as a co-product in the processes of butane or naphtha oxidation to make acetic acid.

Chapter 8

Propylene Derivatives

The usage of propylene in the production of chemicals is indicated by the following tables.

The consumption of propylene for chemical production in the U.S.A. has developed as follows (in '000 long tons).

1950	*1960*	*1965*	*1970*	*1975* (est.)	*1980* (est.)
491	1 110	2 300	3 840	5 800	8 000

The breakdown of consumption in the U.S.A. is as follows:

	1970	*1975* (est.)
Acrylic acid and esters	1%	1%
Acrylonitrile	14%	15%
Cumene	8%	8%
Glycerine	1%	1%
Isopropyl alcohol	21%	14%
Oxo process feedstock	15%	14%
Polypropylene	14%	22%
Propylene oxide	14%	13%
EP rubbers	2%	2%
Miscellaneous	10%	10%

In western Europe propylene usage for chemicals grew from 1.08 million long tons in 1965 to 3.05 million tons in 1970, and a recent forecast suggested that the 1975 demand would be about 5.5 million tons.

The pattern of chemical demand for propylene in Europe is by no means identical with that in the U.S.A. The demand for the Common Market countries as a whole has been assessed as follows:

	1965	*1970*	*1975* (est.)
Polypropylene	7.8%	11.6%	18.3%
Acrylonitrile	2.1%	15.6%	20.0%
Cumene	20.6%	12.8%	10.2%
Isopropyl alcohol	19.8%	11.4%	10.4%
Oxo process feedstock	13.5%	24.6%	18.3%
Propylene oxide	12.7%	12.0%	14.9%
Miscellaneous	23.5%	12.0%	7.9%

The propylene usage statistics for the United Kingdom went somewhat astray in 1970. The 1969 chemical usage of propylene was quoted as 478 000 long tons, and the official statistic so far indicates a 1970 usage of 448 000 long tons. Most market analysts are unconvinced by this and informally put the usage at rather over 500 000 long tons.

It was observed in Chapter 3 that much of the U.S. ethylene production was based on light feedstocks such as ethane and propane. The cracking of ethane yields a mere 0.04 lb propylene per lb ethylene, and propane cracking provides 0.4 lb propylene per lb ethylene. The cracking of a liquid feedstock, such as naphtha, is more flexible in its product ratios. According to cracking conditions, the propylene production per lb ethylene may vary from 0.4 lb for very severe cracking, to little short of 1 lb from cracking of low severity. An average figure for European naphtha cracking could be 0.5 to 0.55 lb propylene per lb of ethylene.

Traditionally a very modest proportion of U.S. propylene has found a chemical use. The main source of propylene is from catalytic cracking which provided over 7.5 million long tons in 1970 in the U.S.A. By 1980 this supply will be over 10 million tons a year of which over 9 million tons will be recovered. Changes in cracking technology could add nearly 2 million tons of extra propylene a year to this total.

Only about 1.3 million tons a year of propylene are obtained as co-product streams from ethylene plants, and the balance is obtained from refineries. At present 60 per cent of total propylene in the U.S.A. is consumed in refineries and 40 per cent in chemical plants. As chemical usage is growing faster than refinery usage the position by 1980 will just about be reversed, with chemicals consuming 60 per cent of the production.

Not only will the chemical demand for propylene grow in the U.S.A., but there is also likely to be a spectacular growth in the supply from chemical (i.e. ethylene) plants. It has been suggested that by 1980 some 8.5 million long tons of propylene a year could be co-produced in ethylene plants in the U.S.A. This arises because of the impending switch to heavier feedstocks in U.S. ethylene plants.

The main refinery application for propylene is in the production of propylene alkylate (made from propylene and isobutane in most cases) as a high octane gasoline component. The value of propylene in this application is the factor that governs the price at which propylene is available to chemical producers in the U.S.A.

The separation of a C_3 gas stream from the gases produced in a catalytic cracker is a relatively simple process, and this gives a stream containing 40–70 per cent propylene, together with propane. For a substantial period this was a suitable feedstock for chemical purposes without further treatment, since in such processes as the manufacture of isopropyl alcohol or cumene the propane is an inactive diluent. With the increasing sophistication of propylene-consuming processes, culminating in polypropylene manufacture, higher standards of propylene purity were demanded. Today the trend is towards standardization on 'polymer grade' (99.9 per cent pure) propylene as the product of commerce.

The European situation has been different in terms of propylene supply. Here it has always been predominantly a co-product with ethylene in naphtha crackers. Refinery propylene is only used to a modest extent in Europe for chemical purposes. There has been no significant production of alkylate gasoline, to foster a large refinery demand for propylene, or to force up the price. Shortages of propylene there have been, but of a temporary nature. The result of this complicated economic pattern is that propylene tends to be lower in price in Europe than in the U.S.A., while ethylene tends to be more expensive in Europe.

In so many different ways the pattern of supply and demand for olefins differs between west Europe and the U.S.A. It comes as quite a surprise to find that according to current forecasts, the total usage of both ethylene and propylene will become very similar in these two great industrial areas during the period 1975–80. The propylene usage will be about half that of ethylene.

Isopropyl Alcohol and its Derivatives

This family of products was one of the first groups of petroleum chemicals to be commercially developed.

Isopropyl alcohol is produced from propylene by hydration. This is most commonly effected indirectly by a concentration-dilution process, involving absorption in sulphuric acid, and subsequent hydrolysis.

$$CH_3CH{=}CH_2 + H_2SO_4 \rightarrow CH_3CH(OSO_3H)CH_3$$
$$CH_3CH(OSO_3H)CH_3 + H_2O \rightarrow CH_3CH(OH)CH_3 + H_2SO_4$$

Operation of the concentration-dilution process, like the production of ethyl alcohol by a similar route, has been established for a considerable number of years. Propylene hydrates more rapidly than ethylene, but, as it also polymerizes more rapidly, by-product formation may present a more serious problem.

The propylene stream will normally be a mixture with propane, containing 40–99 per cent propylene. The liquid phase reaction of propylene with sulphuric acid takes place at temperatures below 40 °C. 20–30 °C is a common range in practice. The sulphuric acid may be of 75–85 per cent concentration, and the pressure is likely to be 20–30 atmospheres. In accordance with the reaction equations indicated, the main initial reaction is to isopropyl acid sulphate. Any higher olefins present will be absorbed as well. The reaction mixture flows to a dilution tank, where water or steam is added to reduce the overall acid concentration to under 40 per cent. The temperature is kept at 35–40 °C.

From the hydrolysis section, the relatively dilute acid/alcohol mixture passes to a steam stripper. The vapours from the top of the stripper, including the alcohol which is volatilized by the steam, are neutralized with sodium hydroxide, and then passed to storage as crude isopropyl alcohol. Dilute sulphuric acid is drawn off the bottom of the stripper, and is passed to an acid concentrator, where its concentration is restored to the required level of 75–85 per cent for re-use.

The crude aqueous isopropyl alcohol is distilled in the first place to give the azeotrope with water, containing 87.7 per cent wt. of alcohol. The recovery of pure isopropyl alcohol is effected by azeotropic distillation, employing entraining agents such as benzene, toluene, xylene, ethylene dichloride or diisopropyl ether.

The process of direct hydration can be applied to propylene as well as ethylene. Such processes have been commercially operated over many years, but until recently the indirect hydration using sulphuric acid was generally preferred. The direct hydration has been operated in both liquid and vapour phase; in one process both liquid and vapour phase reactions were operated together. Metal oxide catalysts were used in the past.

The most favoured operation at present is a vapour phase hydration using a catalyst incorporating phosphoric acid on an inert support. The reaction takes place at 240–260 °C and a pressure of

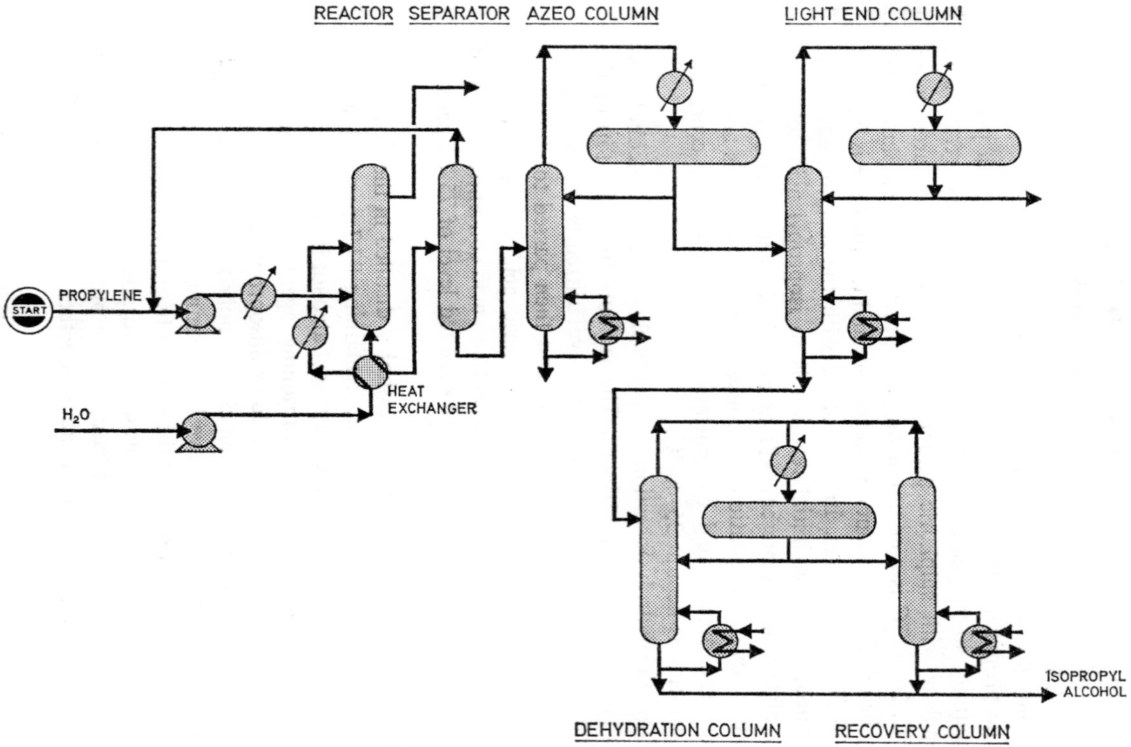

Isopropyl alcohol by direct hydration of propylene

(From *Hydrocarbon Processing*, Nov. 1971)

25–65 atm. The partially reacted mixture is cooled first by heat exchange and secondly by a water cooler. Isopropyl alcohol is scrubbed out of the gases by water washing, and the propylene is recycled to the reaction. The crude isopropyl alcohol is purified by extractive distillation using water as the solvent. The conventional azeotrope of isopropyl alcohol and water is first produced, and dry alcohol may be produced using benzene as an entraining agent.

In a recently announced liquid phase direct hydration, the temperature used is 130–150 °C and the pressure 60–100 atm. The reaction takes place over a catalyst described as a strongly acidic cation exchange resin. A propylene stream (preferably of about 92 per cent purity) passes down a packed tower concurrently with a trickling aqueous phase. The aqueous phase leaving the reactor (if propylene conversion reaches 75 per cent) contains 12–15 per cent isopropyl alcohol. In the purification stage, by-product isopropyl ether is stripped off in the first column, and in the second column the isopropyl alcohol–water azeotrope is produced.

It was once possible to make isopropyl alcohol by hydrogenating fermentation acetone. Now acetone is commonly made from petroleum-based isopropyl alcohol. This gives a vivid illustration of the way in which petroleum-based chemicals introduced new concepts of economic viability in eliminating competitive raw materials.

The main single usage for isopropyl alcohol is still the manufacture of acetone. To make acetone it is normally perfectly feasible to use the 87.7 per cent alcohol–water azeotrope as raw material. Other chemical derivatives include isopropyl esters and hydrogen peroxide. The anhydrous forms of isopropyl alcohol have applications as solvents, gasoline additives, etc.

The applications for isopropyl alcohol in the U.S.A. in 1970 were said to be:

Acetone	47%
Solvent use	20%
Other chemicals	13%
Rubbing alcohol and drugs	12%
Miscellaneous and exports	8%

Production of isopropyl alcohol in the U.S.A. was about 855 000 long tons in 1970. In the United Kingdom the statistics combine both propyl alcohols: in 1970 production amounted to 186 800 long tons, of which the contribution of n-propyl alcohol must have been almost negligible.

The production of isopropyl alcohol has reached a static stage and is now likely to suffer a modest decline. This is due to the fact that alternative methods of acetone production are becoming more significant. In 1970 the proportion of U.S. acetone made from isopropyl alcohol dropped below 54 per cent and this decline is continuing.

Acetone is essentially petroleum-derived, although a classical route was the fermentation process based on starch or molasses, which produced both acetone and n-butyl alcohol. Acetone is a co-product with phenol produced from propylene via cumene (page 134), with hydrogen peroxide by oxidation of isopropyl alcohol (page 128), and with allyl alcohol in one form of glycerine synthesis (page 131). Minor sources of acetone include the paraffin oxidation processes, the oxidation of naphtha to acetic acid and the 'direct oxidation' process to propylene oxide. The cumene-phenol synthesis is by far the most important of these sources.

Vaporized isopropyl alcohol, at 500 °C and about 3 atmospheres, on passing over a brass or copper catalyst, is dehydrogenated to acetone.

A zinc oxide catalyst has also been used for this reaction, in which case a slightly lower temperature, about 380 °C, is used. An alternative liquid phase operation takes place in the presence of an inert high boiling solvent. A Raney nickel catalyst is then used at a temperature of 150 °C.

$$(CH_3)_2CHOH \longrightarrow CH_3COCH_3 + H_2$$

Isopropyl alcohol and acetone are condensed from the reaction gas stream, and separated by fractionation. The hydrogen stream is of value for other chemical reactions, including the frequently associated hydrogenation of mesityl oxide to methyl isobutyl ketone.

There are a number of variations of this widely used process. It can be modified to a combined oxidation-dehydrogenation reaction, passing air with the isopropyl alcohol over a silver or copper catalyst at 400–600 °C. A disadvantage of this route is that no by-product hydrogen stream is formed.

The major importance of acetone, as an industrial product, is indicated by the scale of production. In 1970 demand in the U.S.A.

was about 715 000 long tons and United Kingdom production was 142 200 long tons. The basis of U.S. consumption has been quoted as follows:

Methyl isobutyl ketone and methyl isobutyl carbinol	24%
Methyl methacrylate	18%
Diphenylolpropane (Bisphenol A)	2%
Coatings solvent	9%
Cellulose acetate solvent	14%
Other solvent	4%
Miscellaneous, including exports	29%

Solvent applications for acetone include a variety of uses in the paint and lacquer field, as a dope for cellulose acetate spinning, and as the carrier solvent for acetylene in cylinders.

A chemical application which is still practised is the pyrolysis of acetone to ketene as a step in the production of acetic anhydride.

$$CH_3COCH_3 \rightarrow \underset{\text{ketene}}{CH_2{=}C{=}O} + CH_4$$

The reaction of ketene with acetic acid has been described in Chapter 7.

Perhaps the major group of applications for acetone arises from the series of reactions leading to methyl isobutyl ketone and methyl isobutyl carbinol. These processes are frequently operated as an integrated group of manufacturing units adjacent to the ketone conversion unit.

The first step in this series of reactions is a condensation reaction to diacetone alcohol. Acetone is refrigerated to about 0–5 °C and passed through a column packed with soda lime.

$$\underset{\text{acetone}}{CH_3COCH_3 + CH_3COCH_3} \rightarrow \underset{\text{diacetone alcohol}}{(CH_3)_2C(OH)CH_2COCH_3}$$

Diacetone alcohol has varied applications as a solvent, but its major importance is as a chemical intermediate. An incidental reaction of some commercial significance is its hydrogenation to hexylene glycol. The reaction takes place over a Raney nickel catalyst (this is specially prepared nickel in a finely divided form which is an exceedingly active catalyst for many medium-pressure hydrogenation reactions), at 10–25 atmospheres pressure and 70–80 °C.

Hexylene glycol has a wide variety of industrial applications including its use as a component of hydraulic brake fluids, various solvent applications, action as a coalescing agent in certain co-polymer surface coatings, and use as a de-icing agent for motor gasoline.

$$(CH_3)_2C(OH)CH_2COCH_3 \xrightarrow{+H_2} (CH_3)_2C(OH)CH_2CHOHCH_3$$

diacetone alcohol → hexylene glycol (2-methyl-2,4-pentanediol)

The main stream of diacetone alcohol is directed to the dehydration process forming mesityl oxide.

$$(CH_3)_2C(OH)CH_2COCH_3 \xrightarrow{-H_2O} (CH_3)_2C{=}CHCOCH_3$$

diacetone alcohol → mesityl oxide

The dehydration takes place at 100–120 °C in the presence of an acid such as sulphuric acid. Mesityl oxide has little commercial significance (its unattractive smell being a disadvantage in the development of its solvent applications) except as an intermediate in the production of methyl isobutyl ketone. This hydrogenation reaction takes place at 150–200 °C, and 3–10 atmospheres pressure, over a copper or nickel catalyst. If the hydrogenation is allowed to proceed further, the main product is methyl isobutyl carbinol.

$$(CH_3)_2C{=}CHCOCH_3 \xrightarrow{+H_2} (CH_3)_2CHCH_2COCH_3$$

mesityl oxide → methyl isobutyl ketone

$$(CH_3)_2CHCH_2COCH_3 \xrightarrow{+H_2} (CH_3)_2CHCH_2CH(OH)CH_3$$

methyl isobutyl ketone → methyl isobutyl carbinol

Here one can see one of the benefits accruing from an integrated project, since the hydrogen stream obtained from the dehydrogenation of isopropyl alcohol to acetone may be employed in this hydrogenation.

The two hydrogenation products are separated and purified by distillation. They are important solvents in the paint and lacquer trade. Methyl isobutyl carbinol also finds application in the froth flotation process for the treatment of ores.

Methyl isobutyl ketone reached a peak U.S. usage of 90 000 long tons in 1969, and the 1971 figure was under 85 000 tons. For the future, methyl isobutyl ketone may suffer from being claimed as a pollution hazard, relative to other comparable solvents.

Veba have announced a process which proceeds from acetone to methyl isobutyl ketone in one step. Very few details are available but it would appear to be a simultaneous hydrogenation and condensation, obviously involving two molecules of acetone for each molecule of methyl isobutyl ketone formed.

$$2(CH_3)_2CO + H_2 \rightarrow (CH_3)_2CHCH_2COCH_3 + H_2O$$

This reaction is promoted by an undisclosed catalyst. A moderate temperature of under 100 °C is involved, together with a pressure of up to 80 atmospheres.

There is also said to be a Shell process, which proceeds all the way from isopropyl alcohol to a mixture of acetone and methyl isobutyl ketone in one step, but no details have become available.

The condensation of one molecule of acetone with two of phenol provides diphenylolpropane (otherwise called bisphenol A).

$$2C_6H_5OH + CH_3COCH_3 \longrightarrow HO-C_6H_4-C(CH_3)_2-C_6H_4-OH$$

diphenylolpropane

The reaction takes place in the presence of an acid catalyst, such as hydrogen chloride or sulphuric acid. The operation may be batchwise or continuous. An excess of phenol is used. After separation of the acid and surplus phenol, the diphenylolpropane is purified by distillation and a final recrystallization. Production of diphenylolpropane in the U.S.A. in 1970 was about 85 000 long tons. Some 75 per cent of the total is consumed in epoxy resins (together with epichlorohydrin) and a further 20 per cent goes into polycarbonates (in conjunction with phosgene or diphenyl carbonate). The forecast U.S. market for diphenylolpropane in 1975 approximates to 130 000 long tons. Epoxy resins will continue to be the largest outlet, but polycarbonates are currently expanding at a faster rate.

The first step in the production of methyl methacrylate is the reaction of acetone with hydrogen cyanide, at 25-60 °C and atmospheric pressure in the presence of caustic soda, to form acetone cyanohydrin.

$$CH_3COCH_3 + HCN \rightarrow (CH_3)_2C(OH)CN$$

acetone cyanohydrin

Acetone cyanohydrin reacts with 98 per cent sulphuric acid to give first methacrylamide sulphate, which decomposes in the presence of methyl alcohol to give methyl methacrylate and ammonium hydrogen sulphate. Methyl methacrylate is purified by distillation.

$$(CH_3)_2C(OH)CN + CH_3OH + H_2SO_4 \rightarrow \underset{\text{methyl methacrylate}}{CH_2{=}C(CH_3)COOCH_3} + NH_4HSO_4$$

Demand for methyl methacrylate in the U.S.A. in 1970 was about 200 000 long tons. Polymers of methyl methacrylate are well known in the form of clear sheets used as an alternative to glass in special applications.

Among the other isopropyl alcohol derivatives in the chemical field are isopropyl acetate and the isopropylamines.

The production of isopropyl acetate is a conventional esterification process, using the alcohol and acetic acid on an exactly similar basis to the production of ethyl acetate. Also like ethyl acetate, isopropyl acetate is mainly used as a solvent.

A vapour phase reaction between isopropyl alcohol and ammonia forms a mixture of isopropylamines. These are used as chemical intermediates for dry cleaning aids, herbicides and rubber chemicals.

A specific and rather surprising synthesis is the production of hydrogen peroxide by the oxidation of isopropyl alcohol. This oxidation uses pure oxygen in a liquid phase reaction at 90–140 °C and a pressure of 3–4 atmospheres. A small proportion of hydrogen peroxide must be present as an initiator, and a stabilizer such as potassium pyrophosphate is used. The reaction mixture is diluted with water, and distilled to remove acetone and unreacted isopropyl alcohol. The hydrogen peroxide is recovered as a 6–10 per cent aqueous solution, contaminated with organic impurities. It is concentrated and purified by thermal or chemical methods.

$$\underset{\text{isopropyl alcohol}}{(CH_3)_2CHOH} + O_2 \rightarrow \underset{\text{acetone}}{(CH_3)_2CO} + \underset{\text{hydrogen peroxide}}{H_2O_2}$$

The operation of this process is essentially limited to the U.S.A. Of the U.S. production of hydrogen peroxide, amounting to about 65 000 long tons in 1970, probably less than about 20 per cent is made by this route.

Most hydrogen peroxide is made by an 'autoxidation' route which is, in effect, an indirect oxidation of hydrogen. A substituted anthraquinone can be catalytically hydrogenated to the equivalent anthraquinol. The latter can be oxidized by air back to the quinone form, producing hydrogen peroxide at the same time.

Glycerine

The development of a successful commercial process for the synthesis of glycerine is regarded as one of the major achievements in the petroleum chemical field. Glycerine has long been made as a by-product of the soap and fat-splitting industries. As a by-product, it was subject to wide fluctuations in price and demand. The availability of a synthetic product introduced some stability to the market position.

The first step in the original synthetic glycerine route is the celebrated 'hot chlorination' of propylene. Contrary to the concepts of classical organic chemistry, it is possible, by a choice of reaction conditions, to chlorinate propylene as a substitution reaction, without disruption of the double bond in the propylene molecule.

$$\underset{\text{propylene}}{CH_2{=}CHCH_3} + Cl_2 \longrightarrow \underset{\text{allyl chloride}}{CH_2{=}CHCH_2Cl} + HCl$$

Dry propylene is preheated and mixed with chlorine in a propylene/chlorine mol ratio of 4 : 1. The reaction takes place at about 500 °C and 2 atmospheres pressure, with a contact time of about 2 seconds.

From the cooled reaction products, propylene and allyl chloride are drawn off in a fractionating column. The crude allyl chloride, taken from the bottom of the column, is distilled in two further columns. Pure allyl chloride is taken overhead from the second of these columns. This is the way allyl chloride is made in practice.

An alternative route to allyl chloride has been developed to the small pilot plant stage by Hoechst. This turns the basic reaction into an oxychlorination instead of a chlorination.

$$\underset{\text{propylene}}{CH_2{=}CHCH_3} + HCl + \tfrac{1}{2}O_2 \longrightarrow \underset{\text{allyl chloride}}{CH_2{=}CHCH_2Cl} + H_2O$$

Propylene, hydrogen chloride (which may be in aqueous solution) and oxygen are reacted at 240 °C and atmospheric pressure over a tellurium-bearing catalyst. Corrosion problems may be avoided by the use of acid-resistant materials.

Allyl chloride is reacted with pre-formed hypochlorous acid (formed in a separate tower by reacting chlorine and water) to produce a mixture of dichlorohydrins. The reaction takes place at 28 °C in the aqueous phase. The effluent from the reactor is sent to a separator, from which the aqueous phase is recycled to the hypochlorous acid tower. The non-aqueous phase passes to another reactor, where the dichlorohydrins are converted almost completely to epichlorohydrin by a reaction with milk of lime at below 60 °C.

$$\underset{\text{allyl chloride}}{CH_2{=}CHCH_2Cl} + \underset{\text{hypochlorous acid}}{HOCl} \rightarrow \underset{\text{dichlorohydrin}}{CH_2OHCHClCH_2Cl}$$

$$2CH_2OHCHClCH_2Cl + Ca(OH)_2 \rightarrow 2\underset{\text{epichlorohydrin}}{CH_2{-}CHCH_2Cl \text{ (with O bridging } CH_2 \text{ and } CH)} + CaCl_2.2H_2O$$

The epichlorohydrin is recovered from this operation as a water azeotrope by steam distillation, and concentrated to 98 per cent by further distillation.

Epichlorohydrin is then hydrolysed with a 10 per cent solution of sodium hydroxide in a stirred reactor at 150 °C.

$$\underset{\text{epichlorohydrin}}{CH_2{-}CHCH_2Cl \text{ (with O bridging } CH_2 \text{ and } CH)} + NaOH + H_2O \rightarrow \underset{\text{glycerine}}{CH_2OHCHOHCH_2OH} + NaCl$$

The glycerine is produced as a dilute solution containing salt. This solution is evaporated, desalted and concentrated to 98 per cent by distillation.

Subsequent to the development of this process, the intermediate product, epichlorohydrin, has achieved significance in its own right as one of the raw materials in the range of epoxy resins. These resins have properties of great value in specialized surface coatings and certain electrical and engineering functions. The normal reaction is between epichlorohydrin and diphenylolpropane (itself produced from phenol and acetone).

The synthesis of glycerine by way of allyl chloride, in several variants, has been the main route employed. It has, however, one fundamentally unsatisfactory feature. This is the requirement of chlorine in the initial reaction, only for it to be wasted as unwanted calcium and sodium chlorides in the later stages. More recent efforts have turned, therefore, to the development of a process which would eliminate this need for chlorine. An obvious starting point for such a route would be the oxidation of propylene. Investigations over an extended period have culminated in alternative glycerine processes, commercially operated in the U.S.A.

The first step in the Shell process is the oxidation of propylene to acrolein in the presence of a supported solid copper oxide catalyst. Steam is also added to the reaction, and the temperature requires to be closely controlled, usually at about 350 °C, with a pressure of 2 atmospheres.

$$\underset{\text{propylene}}{CH_2{=}CHCH_3} + O_2 \rightarrow \underset{\text{acrolein}}{CH_2{=}CHCHO} + H_2O$$

Acrolein is a reactive material with a variety of commercial possibilities. In the glycerine synthesis, it is reduced to allyl alcohol by treating it with an excess of isopropyl alcohol. The catalyst used is a mixture of magnesium and zinc oxides, at a temperature of 400 °C.

The counterpart reaction converts the isopropyl alcohol to acetone.

$$\underset{\text{acrolein}}{CH_2{=}CHCHO} + \underset{\text{isopropyl alcohol}}{(CH_3)_2CHOH} \rightarrow \underset{\text{allyl alcohol}}{CH_2{=}CHCH_2OH} + \underset{\text{acetone}}{(CH_3)_2CO}$$

Allyl alcohol is treated with hydrogen peroxide in the presence of a catalyst consisting of tungstic acid or one of its salts. By this means the double bond of the allyl alcohol is hydroxylated to give glycerine. This is a liquid phase reaction carried out at 60–70 °C.

$$\underset{\text{allyl alcohol}}{CH_2{=}CHCH_2OH} + \underset{\text{hydrogen peroxide}}{H_2O_2} \rightarrow \underset{\text{glycerine}}{CH_2OHCHOHCH_2OH}$$

When operating this process on the basis of petroleum raw material, it may be anticipated that the hydrogen peroxide also will be a petroleum chemical, derived by the somewhat unexpected method of partial oxidation of isopropyl alcohol. This process is discussed earlier in this chapter. It may fairly be observed that the obvious advantages of eliminating the chlorine wastage and disposal problem are to some extent offset by the relatively high cost of hydrogen peroxide.

Variations can be effected on some of the steps. The complex, as described, has industrial ramifications extending well beyond the production of glycerine. Acrolein, allyl alcohol, hydrogen peroxide, isopropyl alcohol and acetone are all industrial products with substantial fields of interest outside the glycerine production requirements. The commercial evaluation of such a range of processes is clearly a complicated study.

Another process which has been used in glycerine synthesis is the isomerization of propylene oxide to allyl alcohol in the vapour phase, at 250 °C and using a basic lithium phosphate catalyst.

$$\underset{\text{propylene oxide}}{CH_3CH{-}CH_2\text{ (epoxide, }{-}O{-}\text{ bridge)}} \rightarrow \underset{\text{allyl alcohol}}{CH_2{=}CHCH_2OH}$$

This operation was used in the past by Olin–Mathieson in the U.S.A. as part of a process involving the hypochlorination of allyl alcohol to glycerine monochlorohydrin. This can be hydrolysed to glycerine by a process similar to that used for the hydrolysis of the dichlorohydrin in the conventional synthesis (though the alkali in this case is sodium carbonate).

$$\underset{\text{allyl alcohol}}{CH_2{=}CHCH_2OH} \xrightarrow{HOCl} \underset{\text{glycerine monochlorohydrin}}{CH_2OH{-}CHOH{-}CH_2Cl}$$

Here again one faced the disadvantage of chlorine wastage. More recently the FMC Corporation introduced a process which also starts with the isomerization of propylene oxide to form allyl alcohol. The isomerization step is a liquid phase operation (developed by Progil in France). The propylene oxide is vaporized and bubbled (through a 'sparger') into the reactor liquid. A high boiling solvent is required as the reaction medium, since the reaction takes place at about 275–280 °C. A specially prepared lithium phosphate in powder form is suspended in the reaction medium. The reaction products are vaporized from the reactor liquid and can be separated by fractional distillation.

Allyl alcohol is treated with peracetic acid. This is becoming increasingly used as an oxidizing and epoxidizing agent. Reference to its production is made in Chapter 13.

The reaction of peracetic acid and allyl alcohol provides glycidol and acetic acid. Glycidol hydrolyses to glycerine

$$\underset{\text{allyl alcohol}}{\begin{array}{c} CH_2OH \\ | \\ CH \\ \| \\ CH_2 \end{array}} + \underset{\text{peracetic acid}}{CH_3COOOH} \rightarrow \underset{\text{glycidol}}{\begin{array}{c} CH_2OH \\ | \\ CH \diagdown \\ | \quad\; \rangle O \\ CH_2 \diagup \end{array}} + CH_3COOH$$

$$\xrightarrow{+H_2O} \underset{\text{glycerine}}{\begin{array}{c} CH_2OH \\ | \\ CHOH \\ | \\ CH_2OH \end{array}}$$

The main development of synthetic glycerine has been in the U.S.A. For a number of years now it has accounted for over half the total glycerine production there, but the position seems to have stabilized. In 1970 slightly under 54 per cent of the total U.S. production of 151 000 long tons was synthetic material. The market has been fairly static for some time. In Europe only slightly over one quarter of glycerine production is synthetic material. No synthetic glycerine is made in the United Kingdom. The world position is that some 40 per cent of glycerine is synthetic.

Glycerine usage in the U.S.A. has been quoted as follows:

Alkyd resins	19%
Drugs and cosmetics	18%
Cellophane	13%
Tobacco	12%
Food and beverages	8%
Explosives	4%
Polyurethane polyols	3%
Exports	13%
Miscellaneous	10%

Markets for alkyd resins, drugs and cosmetics grow steadily, and polyurethane polyols represent a rapidly expanding field, but many other outlets are static or declining.

In Europe a higher proportion of total usage (over 50 per cent) is represented by alkyd resins and cellophane. European usage for explosives is also relatively higher than in the U.S.A., but applications in the tobacco, drug and cosmetic fields are smaller.

Phenol

Here we have, once again, a situation in which a series of standard processes, based historically upon coal-derived benzene, has been supplemented by petroleum chemical processes. It is not necessary to dwell on the standard processes, which have been in operation for many years. In essence they are:

(*a*) Sulphonation of benzene, followed by caustic fusion.
(*b*) Chlorination of benzene, followed by hydrolysis with aqueous sodium carbonate.
(*c*) Chlorination, followed by high temperature vapour phase hydrolysis with steam.

The original petroleum route proceeded by way of cumene (isopropyl benzene). Benzene may be alkylated with propylene either in the liquid or vapour phase. The liquid phase process, which once used sulphuric acid as catalyst, but which now uses aluminium chloride at 50–100 °C and at approximately atmospheric pressure, is today rather less in favour. In the vapour phase, benzene may be alkylated with propylene using a supported phosphoric acid catalyst at 250 °C and 25 atmospheres.

$$\underset{\text{benzene}}{C_6H_6} + \underset{\text{propylene}}{CH_2{=}CHCH_3} \rightarrow \underset{\text{cumene}}{C_6H_5CH(CH_3)_2}$$

Purified cumene is fed to an oxidation vessel, where it is emulsified with a dilute sodium carbonate solution. The emulsion is contacted with air and held at 115 °C until 25–30 per cent of the cumene is converted to the hydroperoxide. Carrying the conversion too far involves an increased formation of by-products.

The crude reaction mixture is concentrated to 80 per cent cumene hydroperoxide, and the concentrate fed to an acidifier or 'cleavage reactor'. The peroxide is here treated with dilute sulphuric acid at about 80 °C. One of the means of effecting temperature control is by the evaporation of the water present in the dilute acid—the reaction is strongly exothermic. The pressure in the reactor will normally range between slightly below and slightly above atmospheric. The cumene hydroperoxide is thereby 'cleaved' to phenol and acetone. In this process by-products are numerous. In the oxidation step the most significant by-products are acetophenone and phenyl dimethyl carbinol. By-products of the 'cleavage' include alphamethylstyrene and a number of high boiling components.

$$\underset{\text{cumene}}{C_6H_5CH(CH_3)_2} + O_2 \rightarrow \underset{\text{cumene hydroperoxide}}{C_6H_5C(CH_3)_2OOH}$$

$$\underset{\text{cumene hydroperoxide}}{C_6H_5C(CH_3)_2OOH} \rightarrow \underset{\text{phenol}}{C_6H_5OH} + \underset{\text{acetone}}{(CH_3)_2CO}$$

The reaction products may be separated from each other and from unreacted cumene by distillation or a mixture of distillation and extraction.

Although the cumene route to phenol only became commercially significant in the 1950's, it is today responsible for at least 75 per cent of U.S. capacity, and 70 per cent of west European capacity. It has more recently been supplemented by two new alternative petroleum-based routes to phenol, starting respectively from toluene and cyclohexane. These are discussed in the chapters on aromatics (toluene process) and cyclic compounds (cyclohexane process).

In the U.S.A. the production of synthetic phenol was about 765 000 long tons in 1970, and this was supplemented by a small quantity, perhaps 25 000 long tons, of natural phenol from coal carbonization. These figures would show virtually no increase in the bleak economic climate of 1971, but the overall growth in the U.S.A. is expected to be about 8 per cent a year.

The total demand for phenol in the six Common Market countries and the United Kingdom was expected to reach over 700 000 long tons in 1972 and possibly 850 000 tons in 1975. Of this, a declining proportion is natural phenol, accounting for 7 per cent of production during 1968, and dropping probably below 4 per cent by 1975.

The pattern of phenol demand in the U.S.A. in 1970 was as follows:

Phenolic resins	50%
Caprolactam	20%
Bisphenol A (diphenylolpropane)	10%
Adipic acid	5%
Surface active agents	5%
Miscellaneous	10%

The situation in western Europe is not exactly similar. Considering the totals for the six Common Market countries together with the

United Kingdom, the major application is again phenolic resins, but these only account for about 35 per cent of phenol usage. A further 30 per cent is used for caprolactam, 13 per cent for adipic acid and 22 per cent finds a wide range of other applications.

In two major applications, caprolactam for nylon 6 and adipic acid for nylon 6/6, phenol meets vigorous opposition from cyclohexane as a raw material.

The intermediate common to most caprolactam processes is cyclohexanone. Traditionally the production of cyclohexanone from phenol is carried out in two steps. In the first, phenol is hydrogenated to cyclohexanol at about 150 °C and 15 atm. in the presence of a nickel catalyst. Cyclohexanol is dehydrogenated to cyclohexanone using a zinc-iron catalyst at 400–450 °C. Further processing of cyclohexanone to caprolactam is described in Chapter 13.

Propylene Tetramer (Dodecene)

Propylene tetramer is receding into history. It was a valuable intermediate in the production of anionic synthetic detergents of the alkyl aryl sulphonate type, in which it provided the alkyl radical. The processing was straightforward and the product was excellent in performance but proved to be biologically 'hard'. This means that such detergents, in industrial and domestic waste waters fed to sewage works, were not destroyed by the normal methods of biological degradation. The environmental problems posed included the well-known foaming of rivers and air-blown foam from sewage treatment. The subject is dealt with in general terms in Chapter 17. The net effect is that propylene tetramer (or dodecene) as a raw material for detergents has been substantially reduced in importance. The production level in the U.S.A. is barely half what it was in the mid-1960's, though at over 100 000 tons per annum it can scarcely yet be said to have sunk into insignificance.

The co-product propylene trimer (or nonene) has suffered much the same fate. A major application was in the production of alkyl phenols for detergent use. The alkyl phenols are used with ethylene oxide to form condensation products used as industrial and domestic detergents of the nonionic type. These, too, have proved to be biologically 'hard', and the trend is towards the use of alcohols rather than alkyl phenols in this application.

The production of these lower polymers of propylene is a typical refinery type of operation, and indeed such products have frequently been recovered from refinery 'polymer gasoline' plants. These are in less demand in refineries today, and in some cases are converted to cumene production.

The propylene-containing stream is passed over a supported phosphoric acid catalyst and the reaction conditions are typically 200–240 °C and 15–25 atmospheres pressure.

Propylene tetramer, separated from the reaction products, is used to alkylate benzene at 5–60 °C using a hydrogen fluoride catalyst. The product is called alkyl benzene or detergent alkylate, and these terms are still applied to the present generation of products which carefully use a straight alkyl chain as the alkyl group. Propylene tetramer is a highly branched product and this was found to be the main factor inhibiting biological degradation.

The alkyl benzene is sulphonated and neutralized. The sulphonate is blended with 'builders' such as polyphosphates, silicates, optical brighteners etc. in a soap crutcher. The thick slurry is spray dried in a current of hot air, to form the familiar 'beads' or granules of packaged synthetic detergent.

In most industrially advanced countries there is now either some form of legislation or voluntary agreement to eliminate these biologically 'hard' materials from all domestic detergent products and many industrial applications.

Propylene tetramer has not been made in the United Kingdom since 1964.

Propylene Oxide

The chlorohydrin route is still by far the most important way of making propylene oxide. The chemistry was briefly touched upon in relation to ethylene oxide, but very little ethylene oxide is now made by this means. What has happened, on a substantial scale, is that redundant ethylene oxide units using the chlorohydrin process have been switched to the manufacture of propylene oxide.

To make propylene chlorohydrin, propylene, chlorine and water are introduced into a tower which may or may not be packed. A temperature of around 50 °C is controlled by the flow of raw materials. The effective reactants are hypochlorous acid and propylene.

$$Cl_2 + H_2O \rightleftharpoons HOCl + HCl$$

$$CH_3CH{=}CH_2 + Cl_2 + H_2O \rightarrow \underset{\substack{|\\ CH_2Cl}}{CH_3CHOH} + HCl$$

propylene chlorohydrin

Provided an excess of olefin is maintained, and the concentration of chlorohydrin in the tower is limited to about 6 per cent, relatively little by-product formation of propylene dichloride and other chlorinated products arises from addition reactions with chlorine or hydrogen chloride. The propylene reactions all take place rather more readily than the similar reactions with ethylene, and by-product formation is more evident than in making ethylene chlorohydrin. In some cases the propylene dichloride stream from several such plants may be accumulated and cracked to carbon tetrachloride and perchloroethylene (see Chapter 6—Chlorination of Propane).

The liquid stream incorporating the chlorohydrin is drawn off the top of the tower and treated with an excess of alkali, which may be in the form of a 10 per cent concentration of milk of lime. The mixture is heated and agitated with live steam.

$$2\underset{\substack{|\\ CH_2Cl}}{CH_3CHOH} + Ca(OH)_2 \rightarrow 2CH_3\underset{\substack{\diagdown\diagup\\ O}}{CHCH_2} + CaCl_2 + H_2O$$

propylene chlorohydrin → propylene oxide

Propylene oxide is rapidly removed from the reaction zone. The vapours are condensed and the propylene oxide is separated from water and chlorinated by-products by distillation.

A plant of this type must incorporate special materials of construction to resist the various corrosive influences. The main objection to this process is, of course, the fact that chlorine is used to no ultimate effect, except conceivably to cause a disposal problem.

The same considerations applied to ethylene oxide production, and, after intensive investigation, a direct oxidation process was devised. This intensified the search for a corresponding direct oxidation of propylene. For many years the situation was eased by the ability to use redundant ethylene oxide capacity to make propylene oxide, but, by now, most of this has already either been put to this use or scrapped.

The prospects of finding a catalyst which will enable the reaction of propylene with oxygen to form propylene oxide in the vapour phase in reasonable yield, seem to be slender. In the liquid phase it is possible to use peracetic acid as the oxidizing medium, but a considerable volume of co-product acetic acid is formed as well as propylene oxide. Similarly, in the liquid phase it is possible to oxidize propylene and acetaldehyde together using a cobalt naphthenate catalyst and I.C.I. appear to have taken such a process to the pilot plant stage. Once again acetic acid is obtained as a co-product with propylene oxide.

The first commercial application of what might be termed a direct oxidation process, combines the oxidation of propylene with the dehydrogenation of a hydrocarbon which may, for example, be ethyl benzene or isobutane. With ethyl benzene, the first step is its conversion to the hydroperoxide by air oxidation in the presence of an oxidation initiator. The hydroperoxide reacts with propylene in the liquid phase to form propylene oxide and phenyl methyl carbinol. Reaction conditions are about 90 °C with a pressure of 16 to 65 atmospheres, and a molybdenum naphthenate catalyst is used. The phenyl methyl carbinol is dehydrated to styrene by passing it over a titanium dioxide catalyst in the vapour phase at 180–280 °C. In such a case the production of styrene is likely to be considerably greater than that of propylene oxide, and this obviously limits the general applicability of the process.

$$C_6H_5CH_2CH_3 + O_2 \rightarrow \underset{\text{ethyl benzene hydroperoxide}}{C_6H_5C(OOH)HCH_3}$$

$$C_6H_5C(OOH)HCH_3 + CH_3CH{=}CH_2 \rightarrow \underset{\substack{\text{phenyl methyl}\\ \text{carbinol}}}{C_6H_5CHOHCH_3} + \underset{\text{propylene oxide}}{CH_3\overset{\quad}{CH}CH_2 \ (\text{with O bridging CH and CH}_2)}$$

$$\underset{\substack{\text{phenyl methyl}\\ \text{carbinol}}}{C_6H_5CHOHCH_3} \rightarrow \underset{\text{styrene}}{C_6H_5CH{=}CH_2} + H_2O$$

The intermediate quoted as phenyl methyl carbinol (in accordance with the process announcement) could equally be called phenyl ethyl alcohol, which was an intermediate step in an obsolete route to styrene mentioned in Chapter 7.

The first plant to operate this process was in the U.S.A., isobutane being the saturated hydrocarbon used. There is, of course, a valuable market in the U.S.A. for the isobutylene product, as a raw material for alkylate gasoline. There is a small acetone by-product from this plant. A large plant in Benelux will also use the isobutane version of this process. Since isobutylene does not have much value in Europe, it is apparently proposed to hydrogenate some of this co-product isobutylene back to isobutane for recycle.

A plant to operate on the standard propylene oxide-styrene production basis is to be built in Spain. The original announcement was made a number of years ago and progress seems very slow.

It must also be pointed out that propylene oxide is obtained to a limited extent, accompanied by much greater quantities of other products, in the type of paraffin oxidation practised in the U.S.A., and described in Chapter 6.

The pattern of usage offers some analogies with that of ethylene oxide, but the polyols (or polypropoxy ethers) for application in the preparation of polyurethanes have become the major outlet. The U.S. production of propylene oxide in 1970 was about 535 000 long tons. Of the total usage 42 per cent was for polyols and 27 per cent for propylene glycol. The next biggest outlet is for the higher propylene glycols.

The polypropoxy ethers or polyols are high molecular weight adducts of a polyhydric alcohol (such as glycerine) with propylene oxide.

Propylene glycol itself is a rapidly developing product with a usage in the U.S.A. rising from 200 000 long tons in 1970 to a forecast 320 000 tons in 1975. The major use of propylene glycol in polyester resins accounts for 42 per cent of the U.S. total and a higher proportion still in the United Kingdom. Propylene glycol forms esters with unsaturated dibasic acids and these form an important basis for polyester resin manufacture. Other applications for propylene glycol are in cellophane, tobacco (in the U.S.A.), brake fluids, plasticizers, cosmetics and essences.

Polypropylene

By any normal standard the progress of polypropylene would be regarded as spectacular. Polyolefins have set such extraordinary standards of development that polypropylene is doing no more than follow the precedent set by polyethylene. Up to 1960 the commercial development of polypropylene was slight. The major producer in 1960 was the U.S.A. with about 19 000 long tons. By 1966 U.S. production had reached 249 000 long tons, and demand in the U.S.A. was expected to have doubled by 1970. The figure did not quite reach this level but was not far short at 440 000 long tons.

Demand in west Europe is also rising fast. The figure for 1970 was 295 000 long tons (well over double the 1966 figure) and the forecast demand for 1975 is marginally over 750 000 long tons.

There is substantial international trade in polypropylene, illustrated by the 1970 figures for the United Kingdom. Production was estimated to be 66 000 tons in that year, supplemented by 35 000 tons of imports. Exports were 14 000 tons and consumption was put at 87 000 tons. The pattern of consumption was quoted as follows:

Injection and blow moulding	49%
Fibres (including rope and film yarns)	30%
Sheet and pipe	4%
Film	15%
Miscellaneous	2%

The U.S. pattern also indicates about half the total for injection and blow moulding. The figures for the U.S.A. may be presented in a somewhat different form:

Appliances	5%
Electrical uses	1%
Filament and fibre	27%
Housewares	9%
Luggage	1%
Packaging	15%
Pipe, conduit and fittings	2%
Toys and novelties	5%
Transportation	11%
Miscellaneous	12%
Exports	12%

It has already been pointed out that propylene is a highly desirable monomer for the production of plastics and rubbers, by virtue of its availability on an economic basis. In the case of polypropylene, this general observation has to be qualified by the fact that the usual propylene source (a C_3 stream containing 40–95 per cent of propylene) must be purified to about 99.9 per cent propylene.

The Ziegler-type catalyst system employed in the manufacture of polypropylene to provide a regular molecular structure (described as an isotactic molecule) differs from that used for polyethylene production in that aluminium alkyls are used in conjunction with titanium trichloride rather than titanium tetrachloride.

The reacted catalyst complex in a hydrocarbon medium (comprising, for example, hexane or heptane) is fed to the reactor. The temperature in the reactor may be 50–100 °C and pressures used may be up to 10 atmospheres. By release of pressure, unreacted propylene is removed as a vapour. The polymer slurry is treated (e.g. with alcohol or carbon tetrachloride) to deactivate the catalyst. The polymer is centrifuged away from the solvent, and reslurried or washed with such materials as acetone or isopropyl alcohol. There is a final water wash and the material is then dried. By this means traces of metal salts are eliminated. The polymer is dried, with addition of inhibitor if appropriate, and then extruded and chopped into pellets in the conventional manner.

A random arrangement of propylene monomer units in the molecular chain forms what has been termed an atactic molecule. As has been mentioned, commercial production of polypropylene has concentrated on the regularly arranged isotactic molecular form.

Polypropylene is more rigid and has a higher softening point than any form of polyethylene. It has been found to offer some limitation as a constructional material in its impact strength at low temperatures and it is this fact that has led to the only major process variant in this field. Although there are many claims for modified processes, the production of polypropylene does not have the same wide range of process conditions as does polyethylene. The Ziegler-type catalysts are more or less universally applied. A German process operates the polymerization in the gas phase. The highly purified propylene stream passes into the reactor, in which a quantity of high molecular weight polypropylene powder is kept to maintain the polymerization reaction. A modified Ziegler catalyst is used. The material from the reactor, consisting of gas and solid phases, can be separated simply, and the unreacted propylene may be recycled to the reactor. The proportion of catalyst components present in the polymer produced by this means is so low that it is possible to eliminate catalyst removal or deactivation steps. This type of propylene is claimed to have better impact strength at low temperatures than the conventional product.

Where the product is 100 per cent polypropylene it is described as a homopolymer. It is not uncommon for small proportions of a second monomer to be introduced to make copolymer products which extend the range of properties.

Ethylene-Propylene Rubbers

These have already been considered under ethylene (page 112), but it is obvious that they also represent a significant potential usage of propylene.

C_4 and C_8 Alcohols

n-Butyl and isobutyl alcohols may be produced from propylene by means of the Oxo reaction. This process originated in Germany, and is used industrially in many countries, including the U.S.A. and the United Kingdom. It is a means of extending the chain length of olefins and olefinic compounds.

The Oxo reaction involves the use of synthesis gas to convert olefins to aldehydes, and ultimately alcohols. A description of the process is given in Chapter 11.

Part of the industrial significance of the Oxo process is founded on the feasibility of this route as a means of producing primary alcohols from olefins. The more common petroleum chemical processes normally give secondary alcohols.

Propylene is converted by the Oxo reaction to a mixture of n-butyraldehyde and isobutyraldehyde. As the demand is predominantly for the n-isomer, much attention has been paid to minimizing the proportion of isobutyraldehyde made. In modern processes the ratio of n- to isobutyraldehyde is said to be at least 4:1.

Process modifications (also described in Chapter 11) have enabled the reaction to proceed to the alcohol in one step, and have incorporated a dimerization of the n-butyraldehyde to give a C_8 aldehyde, which may be hydrogenated to give 2-ethylhexanol, $CH_3CH_2CH_2CH_2C(CH_3CH_2)HCH_2OH$.

It is, of course, possible to produce a C_8 alcohol by applying the Oxo reaction to a heptene fraction, but the product called isooctanol, is largely a mixture of dimethylhexanols varying according to the precise nature of the heptene fraction. The heptene fraction is formed by a 'codimerization' or 'oligomerization' process involving propylene and n-butylenes. The process uses a supported phosphoric acid catalyst under pressure, and is very similar to that used in the manufacture of propylene tetramer (q.v., page 137). The heptene fraction is separated by distillation from a heavier gasoline fraction which is produced at the same time.

Until the past few years the bulk of U.S. n-butyl alcohol was made from acetaldehyde. Now probably over 70 per cent is made by the Oxo process. In the United Kingdom all n-butyl alcohol is made by the Oxo process.

Similarly with 2-ethylhexanol; the proportion of U.S. production made from propylene via the Oxo process was about 50 per cent up to the mid-1960's and beyond. Now the proportion must be around 90 per cent. In the United Kingdom, production of 2-ethylhexanol, which had been made from acetaldehyde, ceased in the early 1960's In the past year or two production has recommenced, but this time on the basis of the Shell variant of the Oxo process (see Chapter 11).

Isobutyl alcohol is produced in excessively abundant quantities by the operation of the Oxo process, and by no means all the production can find an industrial outlet. There is also some formation of by-product isobutyl alcohol in the synthesis of methyl alcohol.

In the U.S.A. the production of n-butyl alcohol has been static, and demand actually declined in 1970 to marginally under 180 000 long tons. The pattern of this demand was given as follows:

Solvent use	21%
n-Butyl acetate	14%
Glycol ethers	14%
Plasticizers	12%
Amine resins	10%
Butyl acrylate	7%
Exports	5%
Miscellaneous	17%

A modest growth in U.S. demand for n-butyl alcohol to perhaps 230 000 long tons by 1974 is forecast.

A process for n-butyl alcohol production which stems from Reppe chemistry is operated commercially in Japan. The reaction is:

$$\underset{\text{propylene}}{CH_3CH{=}CH_2} + 3CO + 2H_2O \rightarrow \underset{\text{n-butyl alcohol}}{CH_3CH_2CH_2CH_2OH} + 2CO_2$$

It occurs at 100 °C and 16 atmospheres pressure in the liquid phase. n-Butyl alcohol itself is the solvent medium, and the catalyst comprises a mixture of iron pentacarbonyl, butylpyrrolidine and water. Propylene and carbon monoxide are fed into the catalyst solution, and the formation of n-butyl alcohol commences.

Isobutyl alcohol esters have some applications including plasticizers, and solvents. Usage of isobutyl alcohol is of the order of 30 000 tons annually in the U.S.A.

2-Ethylhexanol continues to grow, more specifically in line with the growth of non-rigid vinyl polymers. The U.S. demand in 1970 was almost 200 000 long tons—an increase of about 50 per cent in five years. A forecast figure for the U.S.A. is about 285 000 long tons by 1974. 2-Ethylhexanol is the basis of the most generally used plasticizer based on C_8 alcohols and phthalic anhydride. In the United Kingdom there are popular products, based on heptenes and on C_6–C_8 wax-cracked olefins, which compete in the field of C_8 plasticizer alcohols. There is also a new development in the production of such alcohols from straight chain olefins which offer certain special characteristics. These C_8 alcohols are all produced by means of the Oxo process.

Of the U.S. demand for 2-ethylhexanol some 80 per cent is for use as plasticizer raw material, 7 per cent goes to make 2-ethylhexyl acrylate, and the remaining 13 per cent goes to export or into miscellaneous uses.

n-Butyraldehyde has some applications as a chemical intermediate apart from the production of n-butyl alcohol and 2-ethylhexanol. The main one is the production of butyric acid, used for cellulose acetate butyrate, and another is the reaction between n-butyraldehyde and formaldehyde to give the chemical intermediate trimethylolpropane.

Acrylonitrile, Acrolein and Acrylic Acid

Acrylonitrile is no longer produced on a substantial scale from acetylene or from ethylene. Major attention is now devoted to the processes for its production from propylene, ammonia and oxygen, described generally as propylene ammoxidation.

Propylene, ammonia and air are reacted, in the presence of a solid catalyst, in the vapour phase. The preferred temperature range is 425–510 °C and the reaction pressure is in the range 1–3 atmospheres.

$$\underset{\text{propylene}}{CH_3CH{=}CH_2} + NH_3 + 1\tfrac{1}{2}O_2 \longrightarrow \underset{\text{acrylonitrile}}{CH_2{=}CHCN} + 3H_2O$$

The original catalyst used was bismuth phosphomolybdate, but other molybdates and metal oxides have been quoted as alternatives. The Sohio process is now understood to be using depleted uranium as a catalyst. This process has been developed to the commercial stage in many countries and variants are quite numerous. The main identifiable difference between the Sohio process and the others is that they largely operate with fixed bed catalysts, whereas Sohio uses a fluid bed.

Acrylonitrile from the ammoxidation reactor is absorbed in water, and purified by azeotropic and conventional distillation.

A variant of the process once operated in the U.S.A. by du Pont is based on nitric oxide, which may be regarded in practice as a reaction product of ammonia and oxygen. This will react with propylene at about 700 °C, in the presence of a supported silver catalyst, to give acrylonitrile.

$$\underset{\text{propylene}}{4CH_3CH{=}CH_2} + 6NO \rightarrow \underset{\text{acrylonitrile}}{4CH_2{=}CHCN} + 3H_2O$$

These reactions do not give very high yields; acetonitrile and hydrogen cyanide are formed as by-products (though they are not always recovered), and a considerable volume of oxides of carbon is also made. The use of a depleted uranium catalyst by Sohio is claimed to improve the yield of acrylonitrile to above 0.8 lb per lb propylene. This figure would represent a yield on propylene marginally above 60 per cent of theoretical. An important further consideration is that the new catalyst also diminishes the yield of by-products, notably hydrogen cyanide.

The development of the propylene ammoxidation processes is a prime example of the successful exploitation of an economical raw material and a single step reaction which may effectively be carried out on a large scale. This is now the dominant method of manufacturing acrylonitrile. The only competitive raw material to have been proposed in recent years has been propane, but this is not likely to be a significant development for the present (see Chapter 6).

By omitting ammonia from the ammoxidation reaction, but otherwise using similar reaction conditions and catalyst, it is possible to make acrolein the major product. With molar ratios of 5 oxygen (as air), 4 steam and 1 propylene at 425–450 °C a yield of 70 per cent acrolein has been claimed, a high figure for this type of operation.

In practice such an operation is frequently taken one step further to acrylic acid. Propylene, in the presence of steam, and oxygen in the form of air, at 400–500 °C over a mixed metal oxide catalyst (oxides of cobalt, molybdenum, bismuth, tin and antimony have been mentioned) is oxidized to acrylic acid. By-products include carbon dioxide, acetic acid and acrolein (which may be recycled).

$$\underset{\text{propylene}}{CH_3CH{=}CH_2} + O_2 \rightarrow \underset{\text{acrolein}}{CH_2{=}CHCHO} + H_2O$$

$$\underset{\text{acrolein}}{CH_2{=}CHCHO} + \tfrac{1}{2}O_2 \rightarrow \underset{\text{acrylic acid}}{CH_2{=}CHCOOH}$$

This operation has been developed to the commercial stage and may well become the dominant route to acrylic acid and acrylates.

There is an alternative propylene oxidation to acrolein used by Shell in the U.S.A. as the first step in glycerine synthesis (see earlier in this chapter).

The development of acrylonitrile in the U.S.A. is demonstrated by production figures:

1954	28 000 long tons
1960	102 000 long tons
1966	318 000 long tons
1970	460 000 long tons
1975 (est.)	760 000 long tons

These figures might be regarded as showing quite a dramatic growth, but the current figures will represent a disappointment compared with the forecasts of a few years ago. The demand for acrylonitrile in the U.S.A. was tending slightly to decline in 1971, following a worldwide lull in synthetic fibre development.

The demand pattern for acrylonitrile in the U.S.A. has been quoted as follows for 1969 and 1975:

	1969	*1975*
Acrylic and modacrylic fibres	48%	55%
ABS and SAN resins	13%	17%
Nitrile rubber	5%	3.5%
Exports	23%	10%
Miscellaneous	11%	14.5%

Other forecasters might show an even more rapid fall-off in exports. In the miscellaneous field the most significant items are the production of Barex resins (copolymers of acrylates and acrylonitrile) and the hydrodimerization of acrylonitrile to adiponitrile.

The position in the United Kingdom is that the 1971 demand for acrylonitrile could be 100 000 long tons for fibres and 20 000–30 000 tons for plastics and rubber outlets. This total demand may increase by about another 30 000 tons by 1973. There are two United Kingdom plants with a total capacity rated at 150 000 tons per annum.

The production of adiponitrile from acrylonitrile has been developed to the commercial stage, and has been much studied in the U.S.A., Europe and Japan. The Monsanto process involves the electrolytic reduction and dimerization of acrylonitrile at the lead cathode of an electrolytic cell. The cathode and anode compartments of the cell are separated by a diaphragm. A vital aspect of this process is that the electrolyte in the cathode compartment is a McKee salt (for instance, tetraethyl ammonium p-toluene sulphonate) which makes it possible to prepare a relatively high concentration of acrylonitrile. Normally acrylonitrile is only soluble in water to the extent of about 7–8 per cent. By means of the McKee salt conductivity is improved and propionitrile formation reduced. This process is a very substantial user of electric power.

There are two variants of the process associated with the Belgian company Union Chimique Belge. One of these, instead of using a cell diaphragm, keeps the anions and cations apart by means of an aqueous emulsion of acrylonitrile. The emulsion is a 2 : 1 mixture of aqueous and organic phases. The aqueous part consists of a dilute solution of potassium phosphate and tetraethylammonium phosphate, with the addition of some sodium hexametaphosphate as a corrosion inhibitor. The organic part is a mixture of fresh acrylonitrile and recycled emulsion. A pilot plant using this process has the mild steel cell lined with polyethylene. Graphite cathodes and iron oxide anodes are employed.

A second variant employs potassium amalgam as the reducing agent. It would seem that sodium amalgam would, in this instance, give rise to excessive polymerization of acrylonitrile and too many side reactions. Potassium is introduced into the system by means of the mercury cell electrolysis of a potassium chloride brine. The potassium leaves the system as potassium sulphate which can be incorporated into compound fertilizers. The hydrodimerization reaction takes place in formamide solvent. This process also has reached the pilot plant stage. It is claimed that the consumption of power is less than with the Monsanto process.

Meanwhile most adiponitrile is made either from adipic acid (originating from cyclohexane or phenol) or from butadiene.

Acetone by the Wacker Process

Reference has already been made to the production of acetone from propylene via isopropyl alcohol, and also to the incidental production of acetone in a variety of other processes. A further interesting prospect is represented by the use of the Wacker process to develop a direct synthesis of acetone from propylene.

The original basis of the Wacker process, as has already been described, was for the oxidation of ethylene to acetaldehyde, using palladium chloride promoted by cuprous chloride. The same principle can be applied to the oxidation of propylene, in which case acetone is the major product.

As the olefin stream will contain a proportion of paraffins, it is normally preferable to operate the process in two steps, with the reaction and oxidation stages kept separate. The reaction conditions in the case of propylene are 90–120 °C and 9–12 atmospheres pressure. The yield of acetone is said to be 92–94 per cent of the theoretical. Crude acetone is purified in a two-column distillation. The first column requires a large number of plates, to effect the almost complete removal of propionaldehyde by-product. This process is commercially operated in Japan.

There are so many by-product and co-product sources of acetone today that this type of process has little opportunity to develop to a significant extent. It represents a useful route to acetone where additional isopropyl alcohol production is not required.

Isoprene

Isoprene has for a number of years been a product of some interest as a minor copolymer with isobutylene in butyl rubber production. As it was required to the extent of only about 2–5 per cent of the monomer mixture, the tonnage involved was relatively slight. Interest in isoprene has been stimulated by the recent development of polyisoprene rubber on the commercial scale. The existing production is commonly based on the dehydrogenation of C_5 olefins present in cracked gasoline fractions. In some cases it has been possible to adapt dehydrogenation plant designed for butadiene production for this purpose. Where it is necessary to develop production of isoprene without reference to existing facilities, alternatives to dehydrogenation techniques have been proposed. One process, that has been taken to the commercial scale, requires, as its first step, the dimerization of propylene to 2-methylpentene-1. The catalyst used is essentially tripropyl aluminium. Propylene is dimerized at 200 °C and 200 atmospheres pressure. The 2-methylpentene-1 is flashed off and purified by distillation. At 200 °C the purified 2-methylpentene-1 is passed over an alkali-modified synthetic zeolite or an acid type catalyst. A substantial degree of isomerization to 2-methylpentene-2 takes place. The reaction mixture proceeds to a pyrolysis furnace, where, in the presence of steam and a hydrogen bromide catalyst, a methane molecule is lost and isoprene is formed.

$$\underset{\text{propylene}}{CH_3CH{=}CH_2 + CH_3CH{=}CH_2} \rightarrow \underset{\text{2-methylpentene-1}}{CH_3CH_2CH_2\overset{\overset{CH_3}{|}}{C}{=}CH_2}$$

$$\downarrow \text{isomerization}$$

$$CH_4 + \underset{\text{isoprene}}{CH_2{=}CH\underset{\underset{CH_3}{|}}{C}{=}CH_2} \leftarrow \underset{\text{2-methylpentene-2}}{CH_3CH_2CH{=}\underset{\underset{CH_3}{|}}{C}{-}CH_3}$$

There is a plant of 75 000 tons per annum capacity in the U.S.A. operating this process. Total production in the U.S.A. of polyisoprene amounted to 122 000 long tons in 1970, an increase of about 10 per cent on the previous year. It has been suggested that demand will grow rapidly in both the U.S.A. and western Europe. The 1971 consumption in the U.S.A. was forecast to reach 125 000 long tons. The same figure could be reached in western Europe by 1975.

The particular significance of polyisoprene is that it is chemically almost identical with natural rubber. Moves towards 'radial ply' tyres have tended to encourage the greater use of both natural rubber and its synthetic counterpart, polyisoprene, in this field. The main difficulty remains the cost of isoprene monomer, which causes polyisoprene to remain a premium product compared with natural

Chloroprene monomer from butadiene

Courtesy Petro-Tex Chemical Corporation

Mitsui phenol/
acetone
plant in Japan

Courtesy
Stone and Webster
Engineering Ltd

rubber and the general purpose synthetics, hence limiting its scale of development.

Ethylene and Butylenes from Propylene

As a means of providing greater flexibility in olefin production Phillips Petroleum Co. introduced their Triolefin process during the 1960's. The reaction involved is a disproportionation reaction whereby two moles of propylene are converted into one mole each of ethylene and n-butylene. The conditions are moderate—temperatures of 120–210 °C and pressures of 25–30 atmospheres have been quoted. The catalyst was of the metal oxide type and could be oxides of molybdenum, cobalt and aluminium in combination. The conversion of propylene per pass is about 40 per cent, and the remaining propylene is recycled. It is desirable to subject the propylene feed to a selective hydrogenation over a palladium catalyst, to minimize the presence of acetylenes and diolefins.

More than 90 per cent of the butylenes produced in this way are usually in the form of n-butylene-2.

Such a process has obvious attractions where surplus propylene is available, and additional ethylene and n-butylenes are required. A commercial plant for this process has been built in Canada.

Chapter 9

Derivatives of C_4 Hydrocarbons

The derivatives of butane, notably from its dehydrogenation and its oxidation, have already been considered in Chapter 6. The raw materials which are to be considered under the present heading are the n-butylenes, isobutylene and butadiene.

Statistics of butylene consumption are rather sparse, but some estimates may be put forward. The usage of all butylenes in the U.S.A. in 1970 was marginally above 2 million tons. Of this total about 80 per cent consisted of n-butylenes and 20 per cent isobutylene. The isobutylene consumption comprised 41 per cent of high purity material for butyl rubber and 59 per cent for other purposes, notably polybutenes.

The usage of n-butylenes was broken down as follows:

Butadiene production	70%
Heptenes	15%
Secondary butyl alcohol	12%
Miscellaneous	3%

It should be noted that the item for heptenes production is quoted gross. Much of the propylene and butylene feedstock emerges as a gasoline fraction, and the butylene figure represents the total feed to this operation. The production of secondary butyl alcohol is almost entirely for further conversion to methyl ethyl ketone.

This pattern of demand is unique to the U.S.A. Elsewhere relatively little of the n-butylenes are dehydrogenated to butadiene. In the near future the whole western European butadiene production will use simple extraction from the C_4 stream arising from naphtha crackers. In western Europe generally the consumption of isobutylene for chemical purposes is comparable with the consumption of n-butylenes. Furthermore, in Europe there is virtually no market for alkylate gasoline, so that there is little difference in the values of butanes and butylenes. In the U.S.A., n-butane and especially isobutane are very inexpensive raw materials, while the value of n-butylenes and isobutylene is enhanced by their value as feedstock, for both chemical synthesis and in alkylate gasoline. Technological developments in the field of catalytic cracking have tended to reduce the availability of butylenes from this source.

Separation of the C_4 Hydrocarbons

Straight fractional distillation is rarely the answer to the requirements of separating the components of a mixture of C_4 hydrocarbons. A careful fractionation will succeed in separating one stream containing isobutane, isobutylene, 1-butylene (one of the n-butylene isomers) and butadiene from another stream containing n-butane and the two 2-butylenes (also n-butylene isomers).

In C_4 streams from refinery catalytic cracking, butadiene is present to a negligible extent, but in the cracking of liquid hydrocarbon streams for ethylene production, butadiene is an important component of the C_4 stream.

The precise nature of the C_4 separation will clearly differ according to individual requirements. Where butadiene is present in the stream, it is common practice to remove this first.

There are a number of possible techniques for butadiene extraction and recovery. The traditional technique, still used, involves the selective absorption of butadiene in ammoniacal cuprous acetate. This is achieved in a multistage extractor/settler system with the hydrocarbon stream flowing countercurrent to the copper ammonium acetate solution. Butadiene is desorbed from the solution by an increase in temperature and a reduction in pressure, and purified by distillation.

More popular today are the techniques of extractive distillation. This takes advantage of the fact that the addition of an external component can affect the relative volatilities of the C_4 hydrocarbons. Extractive distillation using furfural as the external component has been practised for many years, but the most popular extraction medium in recently designed plants has been acetonitrile. More recently still, the virtues of other compounds for this purpose have been canvassed. These include N-methylpyrrolidone, dimethylformamide and dimethylacetamide. Using acetonitrile, where the C_4 mixture is distilled in the presence of the aqueous solvent, butadiene dissolves preferentially and the butadiene/solvent mixture is withdrawn from the bottom of the column. In a second column butadiene is separated from the acetonitrile, and in a third column the final purification is achieved.

An important aspect of this butadiene recovery is the purity of the final product, and in particular the effective removal of traces of acetylenes and unwanted dienes.

The butadiene-free C_4 stream may then be treated for isobutylene removal. With refinery catalytic cracker C_4 streams this becomes the first operation. It is normally achieved by selective absorption in sulphuric acid. The conventional process has used 65 per cent sulphuric acid at 10–20 °C as the absorption medium. Isobutylene may be recovered by diluting the acid to 40–50 per cent concentration and heating. Another common procedure is to warm the 'fat' acid containing absorbed isobutylene to about 80–100 °C for a period, after which isobutylene polymers, about 80 per cent of the total as di-isobutylene, separate out as an upper hydrocarbon layer.

A variant of this process, specifically applied to the recovery of isobutylene, and recently taken to the commercial stage, uses 50 per cent sulphuric acid as the absorbent. Regeneration of the isobutylene is achieved in an elegant operation, in which the recovery of the isobutylene is combined with both dilution and reconcentration of the acid to the required level.

It is possible to recover isobutylene in the hydrated form, as tertiary butyl alcohol, from either of these acid absorption processes. Another process, which would not normally be regarded as a separation process, is the selective polymerization of isobutylene to its polymers, using catalysts such as aluminium chloride, discussed later in this chapter.

The separation of isobutylene from n-butylenes is another instance of the application of the versatile molecular sieve technique. This has been developed by Union Carbide to the pilot plant stage. The specific molecular sieve product is designed to adsorb n-butylenes and to exclude the isobutylene. The n-butylenes are purged from the molecular sieve bed by a higher boiling material which may subsequently be separated from the n-butylenes by distillation. The isobutylene is not obtained by this means in a high degree of purity if butanes are present, since all of the isobutane and most of the n-butane will remain with the isobutylene. It is, of course, possible to purify isobutylene by extractive distillation, but it may be more attractive to pretreat the feedstock by selective hydrogenation and extractive distillation. If the feed to this process consists essentially of isobutylene in admixture with n-butylenes, the separation into pure components can be achieved directly.

After butadiene and isobutylene removal the remaining C_4 stream, comprising the three n-butylene isomers with n-butane and isobutane, does not normally require further separation. Such operations as the hydration of n-butylenes to secondary butyl alcohol can be carried out in the presence of the butanes. In the rare instances where it is required to separate 1-butylene from the cis- and trans-2-butylenes, this may be achieved by distillation.

To separate the butanes from the n-butylenes, extractive distillation using, for example, furfural or acetonitrile as the extraction medium, can be employed. It may be necessary to provide a pure n-butylene feedstock for the process in which the n-butylenes are dehydrogenated to butadiene.

Butadiene from n-Butylenes

The n-butylene stream is recovered by some of the processes outlined above.

Steam and the n-butylene stream are preheated separately, and are mixed in a ratio which may vary from 8:1 to 20:1, according to the concentration of butylenes in the stream and the precise nature of the process used, at a temperature of 600–700 °C. The use of the steam achieves a low partial pressure of the n-butylenes. This favours the reaction equilibrium, and minimizes butadiene polymerization. Moreover, the steam reacts with the carbon deposited on the catalyst, undergoing the water–gas reaction,

$$C + H_2O \rightarrow CO + H_2$$

This eliminates the need, as in butane dehydrogenation, for cycles of air-burning to burn off the carbon deposit. On the other hand, the presence of steam does not permit the use of the conventional dehydrogenation catalysts, such as chromic oxide on alumina. However, such a catalyst can be used for butylene dehydrogenation in a similar way to butane dehydrogenation (see Chapter 6).

In the dehydrogenation reactors the temperature is 625–675 °C, the pressure 0.1 atmosphere and the contact time 0.2 second. The minimum temperature for a reasonable butylene conversion is 600 °C, whilst above 700 °C cracking of butylenes, and polymerization and coking of the butadiene, will occur.

The reaction gases are immediately quenched to about 500 °C, and further cooled. The liquid condensate is charged to a fractionation column and rerun tower, where the C_4 fraction is separated from polymerized material. Butadiene may be extracted from the C_4 fraction by any of the processes outlined in the previous section.

The original dehydrogenation process of this type used ferric oxide as a catalyst, with cupric and potassium oxides as promoters, the whole being supported on magnesium oxide. This catalyst required excessive regeneration and was replaced by a catalyst comprising essentially ferric oxide containing 5 per cent chromic oxide and promoted with a few per cent of potassium oxide. An alternative catalyst now widely used is based on calcium nickel phosphate. While this is more expensive than other catalysts, it offers an improved yield of butadiene, and is the preferred catalyst for butylene dehydrogenation in the U.S.A.

In the U.S.A. the proportion of butadiene derived from the cracking of liquid feeds to ethylene has increased from 15 per cent of the total in the mid-1960's to over 25 per cent today. Of the remainder perhaps 45 per cent comes from butylene dehydrogenation, and rather under 30 per cent from butane dehydrogenation. It is quite commonplace now to forecast that by the early 1980's there will be an adequate supply of butadiene to meet U.S. needs arising from the cracking of liquid feedstocks to ethylene. This situation has just about been reached already in most other parts of the world.

Butadiene by Oxidative Dehydrogenation

Recent studies in Canada indicate that the addition of around 10 per cent of oxygen, to the butylene feed to a dehydrogenation plant, can increase the effective capacity of a butadiene unit based on the calcium nickel phosphate catalyst by as much as 25 per cent. The oxygen reacts selectively with the hydrogen present. This lowers the partial pressure of hydrogen, and also reduces the temperature gradient in the catalyst bed, since the exothermic reaction of oxygen and hydrogen counteracts the effect of the endothermic dehydrogenation. This improves the equilibrium conversion. It is also claimed that carbon deposition on the catalyst is reduced, presumably as a result of a reaction between the carbon deposit and oxygen.

Other than the oxygen addition, the operating conditions recommended were unchanged from the conventional dehydrogenation. The Canadian plant in question used 18 volumes of steam per volume of n-butylene, and a catalyst temperature of 650 °C at the top of the bed dropping to little over 600 °C at the bottom.

The principle of oxidative dehydrogenation of C_4 hydrocarbons to make butadiene has been the subject of much study in the recent past. An ingenious method which was carried to the commercial stage was the Shell IDAS process. The specific reaction here was the removal of hydrogen from butylenes by means of iodine to make hydrogen iodide:

$$\underset{\text{butylene}}{C_4H_8} + I_2 \rightleftharpoons \underset{\text{butadiene}}{C_4H_6} + 2HI$$

The iodine was recovered by carrying out this reaction in the presence of an 'acceptor' which bonds the hydrogen iodide, removing it from the gaseous reaction mixture, and displacing the equilibrium towards maximum dehydrogenation. The 'acceptor' (a metal oxide) with combined hydrogen iodide, was given a regeneration treatment with oxygen and steam. Iodine requires to be recovered with the utmost efficiency.

This process operated in France for a while, but was ultimately abandoned for a number of reasons of process economics, including notably corrosion problems.

There are two processes for making butadiene by oxidative dehydrogenation operated on a substantial scale in the U.S.A. Petro-Tex uses a catalyst involving bromine but, unlike the IDAS process, the halogen is not involved in stoichiometric quantities. Phillips Petroleum have designated their process the O-X-D process. It involves feeding air, steam and n-butylenes over a fixed catalyst bed. No details of the catalyst are given.

A process outlined by BP Chemicals treats n-butylenes at 400–450 °C in the vapour phase with air and steam over a catalyst which may typically comprise oxides of antimony and tin.

Acetic Acid from n-Butylenes

There are no commercial operations using n-butylenes as feedstock to make acetic acid, but there is a considerable patent literature, and two processes of some interest have reached the pilot plant stage. The reason for the interest in the n-butylene feedstock is that it appears in large quantities as a co-product of cracking naphtha to ethylene. In the U.S.A. these n-butylenes, which are mostly derived from catalytic cracking, command a significant value as feedstock for alkylate gasoline, but this is not generally the case elsewhere.

The initial step of the Bayer process is interesting in that it is an example of a reaction between an olefin and an organic acid to give the ester directly. It is normally necessary to convert the olefin to an alcohol first. Bayer carry out a liquid phase reaction between n-butylenes and acetic acid at 110–120 °C and 15–25 atmospheres. The catalyst used is a finely ground solid acidic ion exchange resin, held in suspension. All the n-butylene isomers form the same product, secondary butyl acetate.

$$\underset{\text{n-butylene}}{CH_3CH_2CH{=}CH_2} + CH_3COOH \rightarrow \underset{\text{sec.-butyl acetate}}{CH_3CH_2\underset{|\atop OCOCH_3}{C}HCH_3} \xrightarrow{+2O_2} \underset{\text{acetic acid}}{3CH_3COOH}$$

The second step is also conducted in the liquid phase but without catalysts. The reaction mixture from the first stage is treated with air at 200 °C and 60 atmospheres. Each molecule of secondary butyl acetate breaks down under these conditions into three molecules of acetic acid, one of which is required as recycle to the first stage. From each 0.8 lb of n-butylenes reacted, is produced 1 lb of acetic acid. This means that about 60 per cent of the carbon in the feedstock is converted to acetic acid. A further 28 per cent is lost as oxides of carbon, and some 12 per cent emerges in liquid by-products (formic acid being the most significant) which may be burned.

The operation developed by Hüls is a straightforward gas phase oxidation of n-butylenes with air, over a fixed bed of catalyst comprising titanium and vanadium oxides. The temperature is about 200 °C and a slightly elevated pressure is used. It is important to add steam to depress the explosion limit. The presence of steam is also said to improve selectivity and the yield of acetic acid. On the other hand this means that a very aqueous acetic acid product is obtained, which is costly to concentrate. The reaction is exothermic, so that in practice it can produce some of the steam required by means of waste heat recovery. Formic acid is once again a co-product —to the extent of about 12 per cent of the acetic acid.

Other n-Butylene Derivatives

The hydration of butylenes to secondary butyl alcohol bears a considerable similarity to the production of isopropyl alcohol from propylene by the concentration-dilution process.

The raw material for this operation can be a C_4 gas stream from either a catalytic cracker or a naphtha cracker. It is necessary to remove butadiene (if any is present) and isobutylene. The small proportion of butadiene which may commonly be present in catalytic cracker C_4 streams is selectively hydrogenated to butylene, using a nickel sulphide or copper catalyst. Where butadiene is present in larger quantities, as in the C_4 streams from naphtha cracking, it is necessary to remove it by one of the methods outlined earlier in this chapter. In any case the isobutylene must be removed, and the techniques for doing this have also been discussed.

The C_4 hydrocarbons fed to the acid reaction comprise the butanes and n-butylenes. The butanes pass unchanged through the process, and, as a residual gas fraction, may be returned to an adjacent refinery, or used as fuel within the chemical complex. The liquid phase acid reaction may be carried out using 75–85 per cent sulphuric acid at about 20–35 °C. Particular care must be taken with the reaction conditions, since there is a marked tendency to form polymer by-products. The main product of the acid reaction is butyl hydrogen sulphate. This is diluted with water and heated to hydrolyse it to secondary butyl alcohol. The alcohol is stripped from the dilute acid solution with steam. Sulphuric acid is recovered from the bottom of the strippers, in a relatively dilute form, and concentrated for re-use.

$$\underset{\text{1-butylene}}{CH_3CH_2CH{=}CH_2} + H_2SO_4 \rightarrow \underset{\text{butyl hydrogen sulphate}}{CH_3CH_2CH(OSO_3H)CH_3}$$

$$\underset{\text{2-butylene}}{CH_3CH{=}CHCH_3} + H_2SO_4 \rightarrow \underset{\text{butyl hydrogen sulphate}}{CH_3CH_2CH(OSO_3H)CH_3}$$

$$\underset{\text{butyl hydrogen sulphate}}{CH_3CH_2CH(OSO_3H)CH_3} + H_2O \rightarrow \underset{\text{secondary butyl alcohol}}{CH_3CH_2CHOHCH_3} + H_2SO_4$$

Both 1-butylene and 2-butylene give the same alcohol on hydration. Crude secondary butyl alcohol is taken overhead from the steam stripper, and is concentrated and purified by distillation.

Direct hydration processes have not been successfully applied to butylenes on account of the excessive formation of polymer.

Secondary butyl alcohol has some limited uses as a solvent and as a general chemical intermediate, but some 90 per cent of it is converted to methyl ethyl ketone. The process is similar in principle to the conversion of isopropyl alcohol to acetone. The dehydrogenation reaction takes place in the vapour phase over a catalyst bed of zinc oxide or brass at atmospheric pressure and 350–425 °C. The reaction products are cooled, and the condensed liquids fractionated to separate methyl ethyl ketone. The hydrogen stream is scrubbed to remove any entrained alcohol or ketone, and may then be used for hydrogenation processes elsewhere. It is also possible to produce methyl ethyl ketone from secondary butyl alcohol by a process of catalytic oxidation, but this is not very common in practice.

$$\underset{\text{secondary butyl alcohol}}{CH_3CH_2CHOHCH_3} \rightarrow \underset{\text{methyl ethyl ketone}}{CH_3CH_2COCH_3} + H_2$$

Nearly all methyl ethyl ketone is derived from secondary butyl alcohol, but a small amount is also made from the oxidation of paraffins. This may increase if the butane oxidation route to acetic acid develops.

Methyl ethyl ketone is a low boiling solvent. Its applications are indicated by the breakdown of 1970 consumption in the U.S.A.:

Vinyl coatings	32%
Nitrocellulose coatings	13%
Adhesives	12%
Acrylic coatings	10%
Terephthalic acid manufacture	8%
Miscellaneous coatings	7%
Lubricating oil dewaxing	7%
Miscellaneous uses	6%
Exports	5%

In its applications in the lacquer industry, methyl ethyl ketone finds itself in competition with the ester solvent ethyl acetate. Consumption of methyl ethyl ketone in the U.S.A. was about 200 000 long tons in 1970, and this figure could rise to 250 000 tons in 1975. No figures are quoted for European countries, but the total demand in western Europe is of the order of 80 000–90 000 tons per annum, whereas capacity is substantially above this figure. For the future, methyl ethyl ketone may, in combination with high boiling solvents, be preferred to methyl isobutyl ketone as a lacquer solvent, in view of the alleged pollution hazard of the latter.

As has been pointed out, in western Europe and Japan, the n-butylenes are rather favourably placed as feedstocks for chemical synthesis, since they occur in significant quantities in the C_4 streams from both naphtha crackers and catalytic crackers, while having no major outlet in refinery processing. This has led to renewed interest in the use of the n-butylenes in C_4 streams as feedstock for the manufacture of maleic anhydride. This was earlier used in the U.S.A., but abandoned in favour of a benzene feedstock, which in U.S. terms is a more economical raw material. The economy of the n-butylene feedstock may be offset by a better conversion from benzene.

Both Mitsubishi in Japan and B.A.S.F. in Germany have developed maleic anhydride processes, based on n-butylene, to the commercial stage. A catalyst based on vanadium pentoxide (commonly activated with phosphorus pentoxide) is used in each case, but the B.A.S.F. process uses a fixed bed, and Mitsubishi a fluid bed. The reaction temperature is said to be similar to that used for benzene oxidation—about 350–450 °C. The pressure is likely to be only marginally above atmospheric.

The C_4 stream may be the crude stream from a naphtha cracker, containing butadiene and isobutylene as well as butane and n-butylenes. Although butadiene is converted to maleic anhydride it is normally more economic to extract it from the stream first. If isobutylene is present in the stream it will not provide any maleic anhydride, but will burn and the energy may be recovered in a waste heat boiler. Butane passes through the reactor essentially unchanged.

The appropriate C_4 stream is present with air in the feed to the reactor in a proportion below the lower ignition limit. The maleic anhydride is scrubbed from the reactor effluent by water (or a dilute solution of maleic acid in water). The product, which at this stage is a concentrated solution of maleic acid in water, is dehydrated in an evaporator or similar device, and the crude anhydride so produced is purified by distillation.

Mitsubishi, in particular, have proceeded to use maleic anhydride as a raw material for a series of products more familiar as acetylene derivatives in the field of Reppe chemistry. Reppe reacted acetylene and formaldehyde to produce 1,4-butanediol by way of the equivalent butynediol. The catalytic reduction of maleic anhydride produces first succinic anhydride and then a mixture of gamma-butyrolactone and tetrahydrofuran. The gamma-butyrolactone may be dehydrated to 1,4-butanediol, or used as a raw material for N-methylpyrrolidone—an extraction solvent of increasing importance. The initial step in this three-stage synthesis is to react gamma-butyrolactone with monomethylamine.

Another operation which has achieved a modest commercial development (Hüls have a 12 000 tons per annum plant) is the polymerization of 1-butylene to give the product generally known as polybutene-1. The feedstock may comprise a C_4 stream including olefins and paraffins and containing not less than 50 per cent of 1-butylene. In practice this will have had both butadiene and isobutylene removed by conventional techniques. There is a further pretreatment of the feed to remove water, oxygen, sulphur compounds, acetylenes and dienes. The remaining stream is treated in a stirred autoclave (or pressure vessel) in the presence of a Ziegler catalyst. The catalyst must be extracted from the product. From about 3 tons of C_4 feed, about 1 ton of product is obtained in pellet form. This product has similarities to polypropylene in that it is designed to have a regular molecular structure described as an isotactic molecule. Polybutene-1 is said to lend itself well to the manufacture of pipes, vessels and pumps for handling and storing corrosive liquids at elevated temperatures and pressures. This polymer is claimed to be superior to other polyolefins in the characteristics of stress cracking and cold flow. The original polymer was largely limited to pipe applications, and the increasing range of possibilities has arisen from the development of copolymers incorporating a small proportion of other olefins.

Isobutylene Derivatives

Particularly where naphtha cracking is widely practised, there is normally an ample supply of isobutylene potentially available. In many of its applications it does not require to be separated in the pure form.

Where there is a requirement to produce additional isobutylene, the practice is to dehydrogenate isobutane by similar techniques to those applied to n-butane (see Chapter 6). Also, in Chapter 8, reference was made to the developing process for propylene oxide where the oxidation of propylene is combined with the dehydrogenation of a hydrocarbon (the case was illustrated by the dehydrogenation of ethyl benzene to styrene). If the hydrocarbon used is isobutane, the co-product, with propylene oxide, becomes isobutylene. This represents a route to isobutylene operated on a substantial scale in the U.S.A. where the isobutylene is consumed in alkylate gasoline.

Earlier in this chapter, reference was made to the incidental production of diisobutylene in the course of removing isobutylene from a C_4 gas stream by sulphuric acid absorption. Diisobutylene is used in the production of nonyl alcohol and octyl phenol.

Using a low acid strength and low temperatures, it is alternatively possible to hydrate isobutylene by sulphuric acid to tertiary butyl alcohol. The usual range of conditions is 50–65 per cent sulphuric acid concentration, and a temperature of 10–30 °C.

$$\underset{\text{isobutylene}}{(CH_3)_2C{=}CH_2} + H_2O \longrightarrow \underset{\text{tertiary butyl alcohol}}{(CH_3)_2COHCH_3}$$

The tertiary butyl alcohol is separated from the acid liquor by dilution with water or by solvent extraction. It is of no great importance as a solvent, but is used on a modest scale as a raw material in the production of tertiary butyl phenol, which is an intermediate for oil-soluble phenol-formaldehyde resins.

There is a range of polymers of isobutylene of commercial interest. The lower polymers with molecular weight 300–3 000 are termed polybutenes.

Polybutenes are primarily polymers of isobutylene but some other olefinic and paraffinic material is involved in the commercial product. The feedstock for the polymerization is a C_4 stream (e.g. from a catalytic cracker or naphtha cracker) from which butadiene has been removed. The reaction proceeds in the liquid phase in the presence of aluminium chloride as catalyst and possibly hydrochloric acid as activator. The temperature in the reactor is a major factor influencing molecular weight and viscosity of the product. The polymerization is exothermic and provision of heat transfer facilities, to maintain accurate temperature control, is important. The temperature range used is about 5–50 °C. The pressure is adjusted to maintain a liquid phase of C_4 hydrocarbon, though it is alternatively possible to use a higher molecular weight hydrocarbon, such as hexane, to constitute the liquid phase. The products range from mobile to sticky liquids according to molecular weight.

Consumption in the U.S.A. was 152 000 tons in 1970 and the usages were:

Lubricating oil applications	50%
Caulking and sealing compounds	30%
Adhesives	5%
Rubber compounding	2%
Miscellaneous, including exports	13%

The pattern of demand has not significantly changed in recent years.

Current west European capacity is nearly 70 000 tons per annum, of which about half is in the United Kingdom. In Europe a high proportion of the usage is for lubricating oil applications.

Higher polymers in the molecular weight range 3 000–200 000 are described as polyisobutylenes. The normal polymerization process follows closely the type of operation outlined for polybutenes. A low temperature is employed and catalysts of the Friedel–Crafts type, such as aluminium chloride or boron trifluoride, are required. Usage of these products in western Europe is rather small (less than 10 000 tons per annum). Here again, the major outlet is in lubricating oils where they are used in viscosity index improvers. Other outlets include sheeting, sealants and bituminous type materials.

Isobutylene is a versatile and reactive chemical intermediate, whose full potentialities have probably not yet been developed. At the present time its most important outlet is in the production of butyl rubber.

Butyl rubber is a co-polymer of isobutylene with about 2–5 per cent of isoprene. The co-polymerization is carried out at very low temperature (the feed temperature will usually range between minus 90 and minus 100 °C). The solution of a Friedel–Crafts catalyst, such as aluminium chloride or boron trifluoride, is also injected into the reactors. A large amount of heat is evolved in the polymerization reaction, and this is removed by circulating liquid ethylene through cooling coils (keeping the temperature at about minus 95 °C). A slurry of small rubber particles is formed, and the solvents and unreacted hydrocarbons are flashed off, using hot water. Antioxidants and zinc stearate are added, and after vacuum stripping the rubber crumb is dried, milled and packaged.

Average molecular weight can range quite widely from 200 000 to 800 000, but a typical representation of the polymer molecule, in which the ratio of isobutylene to isoprene molecules is 30 to 1, is as follows:

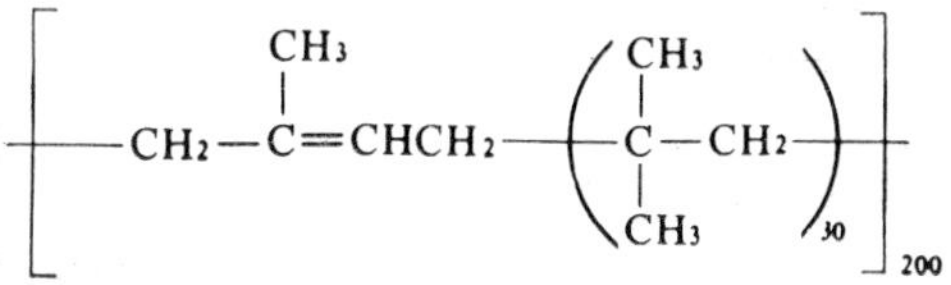

$$\left[-CH_2-\overset{\displaystyle CH_3}{\overset{|}{C}}=CHCH_2-\left(\underset{\displaystyle CH_3}{\underset{|}{\overset{\displaystyle CH_3}{\overset{|}{C}}}}-CH_2- \right)_{30} \right]_{200}$$

This represents a molecular weight of about 350 000.

Butyl rubber has passed through some vicissitudes, since its original market for tyre inner tubes largely disappeared, with the development of the tubeless tyre. The particular merit of butyl rubber is its low permeability to air, and it also possesses resistance to many solvents and chemicals, together with good electrical properties. The essentially saturated nature of the molecule makes the product resistant to ageing (e.g. by air oxidation).

The production of butyl rubber in the U.S.A. in 1970 was 120 000 long tons (representing a significant drop from the 1969 figure). Demand was under 100 000 tons, the balance being taken up by exports. Rather over half the consumption figure is represented by tyres and tyre products. The next most important item is the significant level of exports which may reach up to 30 per cent of the total. Wire and cable products account for some 3 per cent of the total, and the remainder finds a range of miscellaneous uses.

In spite of the arrival of the tubeless tyre butyl rubber has held its own. It is not an ideal material for lining tubeless tyres, in view of its lack of compatibility with other rubbers. One of the areas of development has been the introduction of a small proportion of chlorine into the polymer chain in the form of chlorbutyl rubber.

In some European countries it is still customary to use tyres with inner tubes, but this is likely to be a declining trend. There is a capacity of 127 000 tons per year in western Europe. There is some import from North America, which is rather more than offset by exports, notably to eastern Europe. The current capacity of Esso in the United Kingdom is 44 000 tons per year, but current production in the United Kingdom is significantly less than this.

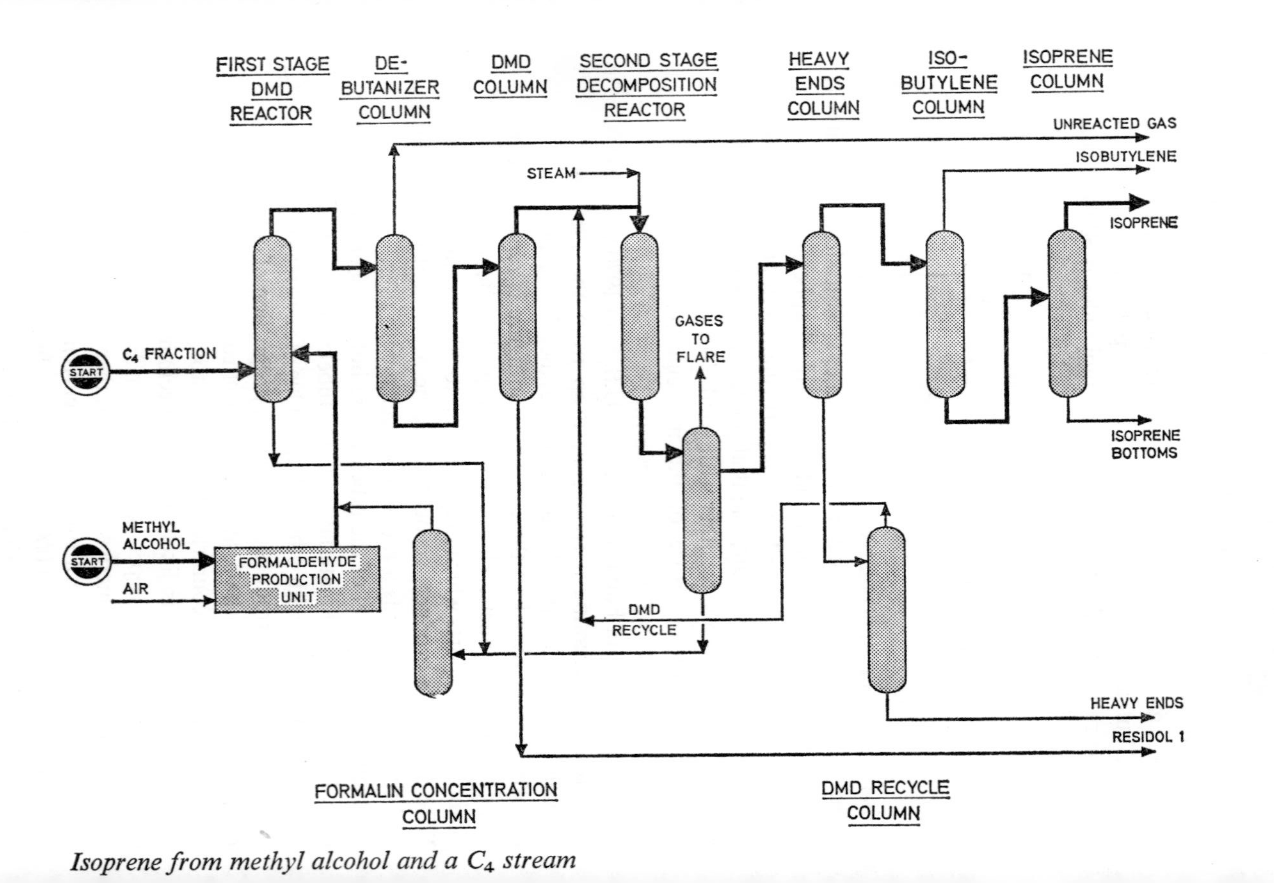

Isoprene from methyl alcohol and a C_4 stream

Butadiene by dehydrogenation of butane and oxidative dehydrogenation of butylenes in one complex in Texas

Courtesy Petro-Tex Chemical Corporation

A corner of
the world's largest
ammonia plant
(Chiba, Japan)

Courtesy
Kellogg
International
Corporation

In some respects butyl rubber would seem to have merit as a general purpose tyre rubber, but some of its characteristics, notably a lack of 'tack' (or bonding power), make it unsuitable for tyre production using conventional methods. Such difficulties can be overcome, but the cost factor then becomes a problem.

There is a process of potential industrial interest employing the Prins reaction between formaldehyde and isobutylene to manufacture isoprene. The process employs two steps. In the first formaldehyde and isobutylene (which may be present in a C_4 stream with butadiene removed) are reacted in the presence of a strong mineral acid catalyst (such as sulphuric acid) at 70–95 °C using a somewhat elevated pressure—up to 20 atmospheres. This forms dimethyldioxane, present in an organic layer which may be separated from the aqueous layer containing unreacted formaldehyde. The unreacted C_4's are separated from the dimethyldioxane which may be further purified by distillation. The dimethyldioxane is then cracked with steam in the presence of a calcium phosphate catalyst at 250–400 °C and a slightly elevated pressure.

The cracking of dimethyldioxane produces isoprene, together with some formaldehyde which is recycled to the first stage reaction.

$$\underset{\text{isobutylene}}{(CH_3)_2C{=}CH_2} + \underset{\text{formaldehyde}}{2HCHO} \rightarrow \underset{\text{dimethyldioxane}}{(CH_3)_2C\langle{}^{CH_2-CH_2}_{O\text{——}CH_2}\rangle O}$$

$$\downarrow$$

$$\underset{\text{isoprene}}{CH_2{=}C(CH_3)CH{=}CH_2} + \underset{\text{formaldehyde}}{HCHO} + H_2O$$

There are several variants of the process and the one which has reached a significant stage of commercial development originated in the Soviet Union. There are several plants of this type in eastern Europe.

Butadiene Derivatives

Butadiene was produced on a very considerable scale during the second world war in the course of the U.S.A. emergency rubber programme. There was a temporary lull in butadiene requirements after 1945, and this period saw the initiation of many investigations into the possibilities of butadiene as a chemical intermediate.

A process which has become of some commercial importance

originates in the chlorination of butadiene to give 1,4-dichloro-2-butylene. The chlorination is here carried out in the vapour phase at 65–75 °C using an equimolar mixture of reactants. The reaction produces a mixture of the 1,4- and 1,2- isomers, but the latter is separated and recycled to the reactor to suppress the formation of additional quantities. 1,4-Dichloro-2-butylene is treated with sodium cyanide or hydrogen cyanide, preferably in the presence of cuprous salts. The 1,4-dicyanobutylene so produced is first hydrogenated in the vapour phase over a palladium catalyst at 300 °C to give adiponitrile. This in turn may be hydrogenated over a supported cobalt catalyst at 100–135 °C and at a very high pressure of around 670 atmospheres to give hexamethylenediamine.

$$\underset{\text{1,4-dichloro-2-butylene}}{CH_2ClCH{=}CHCH_2Cl} \xrightarrow{2NaCN} \underset{\text{1,4-dicyanobutylene}}{CNCH_2CH{=}CHCH_2CN}$$

$$\underset{\text{1,4-dicyanobutylene}}{CNCH_2CH{=}CHCH_2CN} + H_2 \rightarrow \underset{\text{adiponitrile}}{CNCH_2CH_2CH_2CH_2CN}$$

$$\underset{\text{adiponitrile}}{CNCH_2CH_2CH_2CH_2CN} + 4H_2 \rightarrow \underset{\text{hexamethylenediamine}}{NH_2(CH_2)_6NH_2}$$

Alternative routes to adiponitrile operated industrially include the reaction of adipic acid with ammonia, and the hydrodimerization of acrylonitrile. Hexamethylenediamine is one of the two raw materials in the manufacture of nylon 6/6, still the most important form of nylon.

The chlorination of butadiene in the vapour phase at atmospheric pressure and a temperature of 330–420 °C, preferably in the presence of a proportion of butylenes, provides a mixture of 3,4-dichloro-1-butylene and 1,4-dichloro-2-butylene. A proportion of inert gas or diluent should be added to the gas stream. The 1,4-dichloro-2-butylene is isomerized to 3,4-dichloro-1-butylene by heating it to its boiling point in the presence of a catalyst such as cuprous chloride.

3,4-dichloro-1-butylene is dehydrochlorinated by treatment with dilute caustic soda at 85 °C, and the product is 2-chlorobutadiene or chloroprene, the monomer for polychloroprene rubbers. The conventional process for chloroprene manufacture is based on acetylene (Chapter 4). In the past few years the competitive position of butadiene, in relation to acetylene, as a raw material for chloroprene, has continuously improved. New capacity for chloroprene is based on butadiene, and this is rapidly assuming a dominant position in this field.

$$CH_2{=}CHCHClCH_2Cl \xrightarrow{-HCl} CH_2{=}CHCCl{=}CH_2$$

3,4-dichloro-1-butylene — chloroprene

It requires little discernment to detect that it is possible to combine these two operations involving the chlorination of butadiene into one complex from which the ultimate products would be nylon 6/6 and polychloroprene. This operation is now a major source of these materials in the U.S.A., where du Pont will phase out acetylene-based polychloroprene by the beginning of 1974.

Yet another outlet has developed for the 1,4-dichlorobutylene isomer. This is hydrolysed with aqueous caustic soda to produce 1,4-butenediol, and once again we are in the field of Reppe chemistry. The butenediol is hydrogenated to 1,4-butanediol which, according to traditional Reppe techniques, dehydrates to tetrahydrofuran. It is important in this synthesis to minimize the formation of by-product 1,2-butanediol and its polymer.

Work continues both in Japan and the U.S.A. towards the ultimate goal of a synthesis of tetrahydrofuran by a direct oxidation of butadiene, but so far the selectivity and conversion rate have not been acceptable.

As might be expected, consideration has also been given to the production of adiponitrile from butadiene without the use of chlorine. There is a potential process, which does not appear to have been commercially developed, in which hydrogen cyanide is directly reacted with butadiene. The mixture of unsaturated mononitriles so formed is separated and isomerized to form essentially linear pentene nitriles. These nitriles are reacted with more hydrogen cyanide to form a mixture of dinitriles, from which adiponitrile is isolated and purified. A nickel catalyst can be used in all three steps, and patent disclosures cite other metals also.

Another proposed process of some relevance in this field proceeds in several steps from butadiene to hexamethylene diamine. The significance of adiponitrile in the nylon field is that it hydrogenates to hexamethylene diamine. Butadiene will react with iodine and copper cyanide to produce the cuprous iodide complex of dehydroadiponitrile. This complex may be hydrolysed by aqueous hydrogen cyanide to dehydroadiponitrile. The latter may be readily hydrogenated to hexamethylene diamine. This process has not been commercially proven.

Another process, which is operated on a limited commercial scale, is the treatment of butadiene with sulphur dioxide, under conditions of raised temperature and pressure. This reaction forms a sulphone, which, on hydrogenation, yields tetramethylene sulphone, known commercially as sulfolane.

$$\underset{\text{butadiene}}{CH_2{=}CHCH{=}CH_2} + SO_2 \rightarrow \begin{array}{c} CH{=}CH \\ | \quad\;\; | \\ CH_2 \;\; CH_2 \\ \diagdown\;\diagup \\ SO_2 \end{array} \xrightarrow{+H_3} \begin{array}{c} CH_2{-}CH_2 \\ | \qquad\; | \\ CH_2 \;\; CH_2 \\ \diagdown\;\diagup \\ SO_2 \\ \text{sulfolane} \end{array}$$

Sulfolane first attracted attention as a selective solvent for aromatics, and it has been used in the place of aqueous diethylene glycol in processes for aromatics separation. The Udex process, which is the most widely used process of this type, can be adapted to the use of either solvent medium. It is frequently possible to increase the capacity of an existing unit of this type, when switching to sulfolane extraction, by taking advantage of the low solvent to feed ratio which becomes possible.

Processes known as the Adip and Sulfinol processes are modifications of the standard alkanolamine extraction of acid gases, such as hydrogen sulphide and carbon dioxide, from a gas stream. In the Sulfinol process the absorption medium is normally a mixture of sulfolane, diisopropanolamine and water, rather than the conventional aqueous solution of alkanolamine. The modified process is claimed to minimize utility costs, and to widen the range of gases which can be effectively treated. The process is applicable to the treatment of sour natural gas, or the purification of synthesis gas.

An interesting derivative of butadiene is the cyclic trimer, 1,5,9-cyclododecatriene. This is obtained once again by the use of Ziegler catalysts. A modification, developed by Mitsubishi in Japan, uses a combination of a polyalkyltitanate and dialkylaluminium monochloride as catalyst. This enables the trimerization of butadiene to proceed directly from a C_4 stream, without the need to separate it first. The butanes and butylenes in the stream are virtually unaffected.

The C_{12} ring compound so obtained may be hydrogenated to the C_{12} cycloparaffin, cyclododecane. This is subjected to a series of reactions exactly analogous to those by which caprolactam is obtained from cyclohexane, and the ultimate product is lauryl lactam.

The German operation of Hüls follows the conventional pattern of reaction to cyclododecanone, then to its oxime derivative, and thence by the Beckmann rearrangement to lauryl lactam. This chemistry is dealt with in some detail in Chapter 13. The Japanese technique operated by Toray (perhaps better known under its earlier name of Toyo Rayon) again follows the national precedent. The specific 'photonitrosation' technique (see Chapter 13) forms cyclododecanone oxime hydrochloride in one step from cyclododecane. The Japanese process to lauryl lactam, therefore, has several fewer steps than the original German operation.

Lauryl lactam is the monomer for nylon 12. This is not a fibre material, but a specialized resin with some specific value arising from its impermeability to moisture. It is employed in the form of pipe and in 'powder coating'.

Apart from the chemical applications of butadiene, it has increasing significance as a copolymer in the production of latices used, for example, in surface coatings, and in the important copolymer resins designated ABS (acrylonitrile–butadiene–styrene).

Although due attention must be paid to the chemical outlets for butadiene, it is the usage in synthetic rubber that accounts for most of the total. In the U.S.A. in 1971 the domestic consumption of butadiene was about 1.4 million long tons. The usage was as follows:

Styrene-butadiene rubber and latex	58%
Polybutadiene rubber	17%
Nitrile rubber	2·5%
Polychloroprene	3%
Nylon 6/6	11%
ABS resins	4·5%
High styrene-butadiene copolymer resins	4%

The rubber outlets account for about 80 per cent of the total, as can be seen. The United Kingdom production of butadiene in 1970 was 173 000 long tons, and over 90 per cent of this found outlets in synthetic rubber manufacture.

In western Europe as a whole there is at present a reasonable balance between supply and demand for butadiene. The production, very largely from extraction of naphtha cracker C_4 streams, amounted to about 780 000 long tons in 1970 and demand was marginally less. Over 90 per cent of this butadiene consumption was for synthetic rubber of varying kinds.

The usage of butadiene in Japan was 513 000 tons in 1971 and this figure is expected to rise to almost 700 000 tons in 1975. Here again well over 90 per cent of the total is for rubber. In 1971 rather over 60 per cent of total Japanese usage was for styrene–butadiene rubber. By 1975 the proportion is estimated to drop slightly to 58 per cent. In Japan there is still substantial growth taking place in styrene–butadiene rubber production.

Future expansion of butadiene consumption is limited by the restricted growth of the general purpose rubbers. Styrene–butadiene rubber is, in general, fairly static, though there is significant growth in latices. Nitrile rubber is also growing very slowly. Polybutadiene rubber is the fastest moving of this group. In the U.S.A. there is, for the time being, a rapid growth in butadiene consumption for polychloroprene rubbers, but this will slow down once the replacement of acetylene in this field is complete. The butadiene-based nylon materials appear to be more than holding their own in the U.S.A., and the ABS resins represent another growth feature. In Europe there is no butadiene-based nylon and relatively little butadiene-based polychloroprene. There are possible areas of expansion here. As European butadiene is essentially a co-product of ethylene, it is clearly desirable that, as far as possible, butadiene consumption should keep pace with that of ethylene. Alternatively it would be necessary to develop ethylene production from natural gas liquids, so that butadiene co-production would be minimized.

SBR is the designation of the most common form of synthetic rubber, derived from the copolymerization of butadiene and styrene in the approximate proportion of 3 to 1 by weight. The polymerization may be effected as an emulsion process in water. Amongst the additional components, a catalyst is added to promote the polymerization and a stabilizer to keep the rubber particles in suspension. Some heat is applied to start the reaction, but subsequent cooling is necessary since the reaction is exothermic. It is required to maintain a fairly low temperature (about 5 °C) in the polymerization vessel, since the 'cold rubber' produced under these conditions has the best physical properties. The latex obtained is stripped of unreacted material, and coagulated to 'crumbs' which are dried and baled.

A further processing development for SBR has been the introduction of solution polymerization. The copolymerization proceeds in hydrocarbon solution in the presence of organometallic complex catalysts of the Ziegler type, discussed in connection with high density polyethylene. This development operates on a relatively modest scale, but it is claimed to offer more control of the molecular structure of the product than the conventional process.

SBR is now the major component of tyres for ordinary passenger cars, although it is not suitable for the larger and heavier truck or aeroplane tyres, owing to its liability to overheating when so used.

By increasing the proportion of styrene in the copolymer, speciality rubbers may be developed for compounding to impart higher resistance to abrasion.

SBR is still by far the major synthetic rubber in scale of production. In the U.S.A. in 1970 over 1.1 million tons of SBR were produced, though this represents a slight drop compared with the previous year, and a fairly static level of production for 4–5 years. It should be noted that figures some 20 per cent or so higher would be given if the oil extenders are counted in. The figure quoted is for rubber only. Although SBR is primarily a tyre rubber, this is by no means its only use. The proportion used for tyres and tyre products in the U.S.A. is now slightly lower than the traditional 67 per cent, probably about 60 per cent. A further significant percentage has been exported but this has dropped with the construction of production facilities in many countries. Over 30 per cent of the total finds miscellaneous uses for mechanical goods, shoes, foams, wire and cable coating and many others.

The production in the United Kingdom of SBR as rubber and latex was almost 200 000 long tons in 1971.

The advent of the stereospecific rubbers (notably polybutadiene) has robbed SBR of much of its growth potential, since the new products are directly competitive in a number of cases.

NBR represents the nitrile rubbers, which are produced by the copolymerization of butadiene and acrylonitrile. The technique of manufacture is essentially similar to the emulsion polymerization process for SBR. These are specialized rubbers, owing their significance to an exceptional degree of resistance to hydrocarbon solvents, more particularly the aromatics.

The proportion of the nitrile rubber monomers approximates to 67 per cent butadiene and 33 per cent acrylonitrile. The production of nitrile rubber in the U.S.A. in 1970 was 68 000 long tons. The U.S. figure has remained almost static for some years. Production in western Europe is comparable with that in the U.S.A., being of the order of 60 000 to 65 000 long tons in 1971. A steady, if unspectacular, increase is forecast for west European nitrile rubber. The production in 1976 should be about 50 per cent higher than in 1971.

The polymerization of butadiene directly to form polybutadiene rubber (BR) has achieved considerable prominence in the past few years. This is evidenced by the fact that the U.S. production of polybutadiene, which was 202 000 long tons in 1967, reached 284 000 long tons in 1970. It should more than double this figure by 1980. The production of polybutadiene in western Europe was approaching 200 000 tons in 1971.

It is possible to produce polybutadiene by an emulsion polymerization process at fairly low temperatures in the region of 5–35 °C in an aqueous medium. This requires the use of modifiers (tertiary mercaptans) for molecular weight regulation, soaps to effect the emulsification, an organic hydroperoxide as polymerization initiator and an activator for the hydroperoxide. The molecular configuration from such a process is mixed, including only about 14 per cent of the cis-1,4-structure.

The more common processes for polymerization of butadiene are based on solution polymerization. The catalyst used in one type of operation is an alkyl–lithium such as butyl–lithium. The polymer produced then has about 35 per cent cis-1,4-structure. Alternatively a Ziegler catalyst (e.g. triethylaluminium and titanium tetrachloride) may be employed, in which case the cis-1,4-proportion in the final product may rise to 97 per cent. Operating conditions may vary widely according to individual circumstances; the temperature at which the polymerization takes place, for instance, may vary from 10 to 120 °C. Where the polymer takes the cis-1,4-structure, its properties become closely akin to those of natural rubber.

```
H         H   H         H
 \       /     \       /
  C===C         C===C
 /       \     /       \       /
CH2       CH2—CH2       CH2—CH2
```

An argument rages as to the relative merits of the cis-1,4-structure. The processing characteristics of polybutadiene are not yet all that is desirable, and it has been claimed that this is a particular difficulty with products with a ‘high cis’ content. It is common practice to blend polybutadiene with other rubbers, such as SBR or natural rubber, to assist processing and to improve skid resistance of the tyre tread on wet roads. It has been argued, on the other hand, that 'high cis' polybutadiene has superior properties. Certainly the bulk of the production at the present time is of the 'high cis' type. This is not true of the United Kingdom where the alkyl–lithium catalyst system is in use at the one existing plant.

Among process variants one may note the Phillips Petroleum Solprene process, which polymerizes butadiene direct from the C_4 stream, thus obviating the butadiene separation and purification steps. Goodyear have been credited in the past with a process of their own using a cobalt catalyst, but a more recent announcement suggested that their new unit at Beaumont, Texas, employs a nickel-based catalyst.

Chapter 10

Derivatives of Higher Olefins

Reference has already been made to the possibility of producing isoprene from propylene. The specific interest of isoprene arises from the development of polyisoprene rubbers, which are competitive, not with other synthetic rubbers such as SBR, but with natural rubber. The economic requirement is that the isoprene monomer must be produced at a high purity, on a basis which will permit the polymer to compete effectively with natural rubber.

One process in operation for isoprene manufacture in the U.S.A. and the Netherlands is the dehydrogenation of isopentenes (particularly 2-methyl-2-butylene) in equipment closely similar to that used for the dehydrogenation of butylenes to butadiene. This depends upon an appropriate source of C_5 unsaturates, which can be separated in some cases from a gasoline stream. It is possible to concentrate the isoprene precursors from a C_5 gasoline stream by an extraction process using 65 per cent sulphuric acid.

A process is available for the dehydrogenation of isopentane over a chromia-alumina catalyst, and this is a close counterpart of the corresponding process for dehydrogenating n-butane to butadiene. It is possible to use a mixed isopentene–isopentane feedstock for this operation.

In such dehydrogenation processes the main separation step is likely to involve extractive distillation, using the same set of solvents as are used in the separation of butadiene by this means.

Where naphtha cracking is carried out, the C_5 fraction of the cracked spirit contains a significant proportion of isoprene. The main difficulty here lies in the large volumes to be fractionated to recover a relatively small component. The separation of isoprene from the other C_5 components is not difficult, but it is necessary to include provision for the removal of cyclopentadiene and the separation of piperylene.

There is no shortage of other processes mooted for isoprene production. For the present, however, the development of polyisoprene remains on a modest scale, so that the total number of manufacturing facilities is small.

The main current significance of C_6 olefins relates to the propylene dimers. Both 2-methylpentene-1 and 2-methylpentene-2 represent steps in the production of isoprene from propylene (see Chapter 8).

The propylene dimer 4-methylpentene-1 has recently achieved some significance as a monomer for the production of a series of synthetic resins with some glass-like properties, designated TPX resins by I.C.I.

Hexene-1 has recently assumed some interest as a co-monomer with ethylene in the production of polymers with properties similar to high density polyethylene.

The commercial products known as heptenes are in fact complex mixtures of heptenes, methyl hexenes and dimethylpentenes. Such a mixture is made from propylene and butylenes using a phosphoric acid catalyst as outlined in Chapter 8. As with many of the higher olefins, the main significance of heptenes is as an olefinic feedstock to the Oxo reaction, aimed at the production of a primary alcohol. The alcohol made from heptenes is a mixture of C_8 alcohols generally called isooctanol, and used in plasticizers for polyvinyl chloride.

The recovery of diisobutylene as a possible step in the purification of a C_4 stream was mentioned in Chapter 9. Diisobutylene has achieved some significance as a raw material for the Oxo (or carbonylation) process where this C_8 olefin is converted into the C_9 alcohol, nonanol (or nonyl alcohol). Nonanol is used in the production of phthalate esters for PVC plasticizers. Diisobutylene is also a component of octyl phenol which has been used in lubricating oil additives and in nonionic detergents.

The Oxo process may be applied to other long-chain olefins, converting them to the corresponding primary alcohol with one extra carbon atom. Olefins in the C_6–C_8 range, either separately or as distillation fractions, may be converted to the C_7–C_9 primary alcohols. The main market for such alcohols is again in the production of their esters as plasticizers.

A recent trend in the production of the higher olefins has emphasized the straight chain structure. The production of such straight chain olefins by the controlled polymerization of ethylene has already been mentioned in Chapter 7.

The alternative methods are the cracking or dehydrogenation of straight chain paraffins. The wax or other paraffinic feedstock is treated by means of the molecular sieve or urea adduct processes described in Chapter 6. The straight chain paraffins are then cracked or dehydrogenated (thermally or catalytically) using conditions designed to retain the straight chain structure, and to maximize the double bonds in the terminal or alpha position. The straight chain alpha olefins obtained by such means can never compare in purity with those from ethylene polymerization, but they are lower in cost.

Straight chain olefins in the C_7–C_9 range find most applications in the field of plasticizer production. The olefin is converted to the primary alcohol (which will be in the C_8–C_{10} range) and this is esterified to form the plasticizer. Such plasticizers have benefits and limitations compared to the conventional materials. The electrical properties are relatively poor, so that they are not suitable materials for cable compounds. They do impart superior low temperature flexibility, however, and may obviate the need to introduce more expensive esters, such as adipates, into the formulation to impart this property.

Straight chain olefins in the C_{11}–C_{14} range can form the alkyl chain in the production of linear alkyl benzenes. These are intermediates for much of the current manufacture of biodegradable detergent materials. Although there is a considerable scale of production of linear alkyl benzenes based on the conventional alkylation of benzene with straight chain olefins, using hydrofluoric acid as alkylation catalyst, it is also possible to proceed directly from the n-paraffin using first a monochlorination step, and, secondly, an aluminium chloride alkylation catalyst.

The use of straight chain higher olefins to make alcohols by the Oxo reaction has already been mentioned. There is a range of possibilities for these materials in the detergent field. Primary alcohols in the C_{12}–C_{18} range may be sulphated directly to form the traditional primary alkyl sulphate detergents. Primary alcohols in the C_{10}–C_{18} range can first be ethoxylated with an appropriate proportion of ethylene oxide and the resultant product sulphated and neutralized. This gives rise to the ether sulphates, another form of anionic detergent. Alternatively, alcohols in the C_{16}–C_{18} range may be ethoxylated to give low-foaming nonionic detergents. These indications merely typify the range of possibilities in this field of biodegradable detergent production. Also competitive in this area are conventional fatty alcohols produced by catalytic hydrogenation of fatty acids, straight chain alcohols produced directly from ethylene polymerization (see Chapter 7), and the secondary alcohol ethoxylates in which the secondary alcohols are obtained by the oxidation of n-paraffins.

Still in the detergent field, the straight chain C_{15}–C_{18} olefins may be sulphonated directly to form alkenyl sulphonates. These sulphonates retain the double bond in the molecule and therefore are obtained using a very mild sulphonation technique. It is of some particular importance in this production that the olefin feedstock should, so far as possible, consist of alpha olefins. These have now been products of promise rather than performance for a number of years, but they are made on a limited commercial scale.

The once-prominent sulphation of olefins to form secondary alkyl sulphates is still carried out on a limited scale. Such products are most effective as components of detergent blends. The secondary alkyl sulphates, used on their own as the active material for a detergent formulation, were found to be inadequate in foaming power and in storage characteristics.

It is apparent that most of the applications for the C_{10} and higher olefins lie in the detergent field. Such olefins also have some applications as chemical intermediates for textile finishes, thickeners, cosmetic ingredients, synthetic waxes and petroleum additives.

B. PRODUCTS FROM MISCELLANEOUS PETROLEUM SOURCES

Chapter 11

Derivatives of Synthesis Gas

Synthesis gas is the common designation for a mixture of hydrogen and carbon monoxide which may form the basis for the synthesis of ammonia, methyl alcohol and for the Oxo reaction.

The production of synthesis gas did not start with petroleum chemicals. The search for a process for the fixation of nitrogen, which culminated in the production and application of synthesis gas in the manufacture of synthetic ammonia, is commonly regarded as one of the most stirring episodes in the development of the chemical industry prior to 1914. The original production was based on coke.

Ammonia and its Derivatives

It is not proposed to attempt any full account of ammonia production, but to give some indication of the influence of petroleum raw materials on an already flourishing large-scale industry.

The basic coke–water-gas reaction is as follows:

$$C + H_2O \longrightarrow CO + H_2$$

Each molecule of carbon monoxide is replaced by a molecule of hydrogen using the shift reaction. (In earlier terminology this was sometimes called the water-gas equilibrium.)

$$CO + H_2O \rightleftharpoons CO_2 + H_2$$

It is also required to purify the hydrogen and to introduce nitrogen into the mixed gas stream. Using coke as a raw material the nitrogen is commonly introduced at an early stage, during the blow run of the water-gas generators (i.e. that portion of the cycle which builds up the temperature in the fuel bed). The mixture of nitrogen and hydrogen, appropriately purified, is reacted at high pressure in contact with a catalyst to complete the deceptively simple equation,

$$N_2 + 3H_2 \rightleftharpoons 2NH_3$$

This, in principle, was the basis of ammonia production from coke before the advent of petroleum chemicals. The first production of ammonia from natural gas dates back to 1930.

The original process used to recover hydrogen from natural gas was a thermal cracking process:

$$CH_4 \rightarrow C + 2H_2$$

The hydrogen stream from such an operation is more concentrated than with most other processes and the purification techniques may therefore be less elaborate. The most successful basis of purification is low temperature separation. The carbon black also obtained from this cracking process (a type known as a 'thermal' black) has certain uses in rubber and surface coatings. The application of this particular process did not develop, since it is less economical in the use of natural gas than subsequent alternative processes, and it has now been abandoned.

The next development in the use of natural gas dates back to the early 1930's with the advent of the methane–steam reaction

$$CH_4 + H_2O \rightarrow CO + 3H_2$$

A comparison of this equation with the coke–water-gas reaction will illustrate clearly that this is a more favourable basis for the production of hydrogen.

A similar type of reaction can be drawn up for any hydrocarbon, and techniques have in fact been devised to apply such a reaction to a wide range of products, from methane up to fractions in the gasoline range. The hydrogen present in the hydrocarbon assists in the production of a synthesis gas containing appreciably more hydrogen than carbon monoxide. Whilst the shift reaction will subsequently provide a molecule of hydrogen for each molecule of carbon monoxide converted, it is economically favourable to start with the maximum hydrogen content in the crude synthesis gas. This ensures that the production of carbon dioxide in the shift reaction is minimized, and corresponding savings are made in the cost of its ultimate removal.

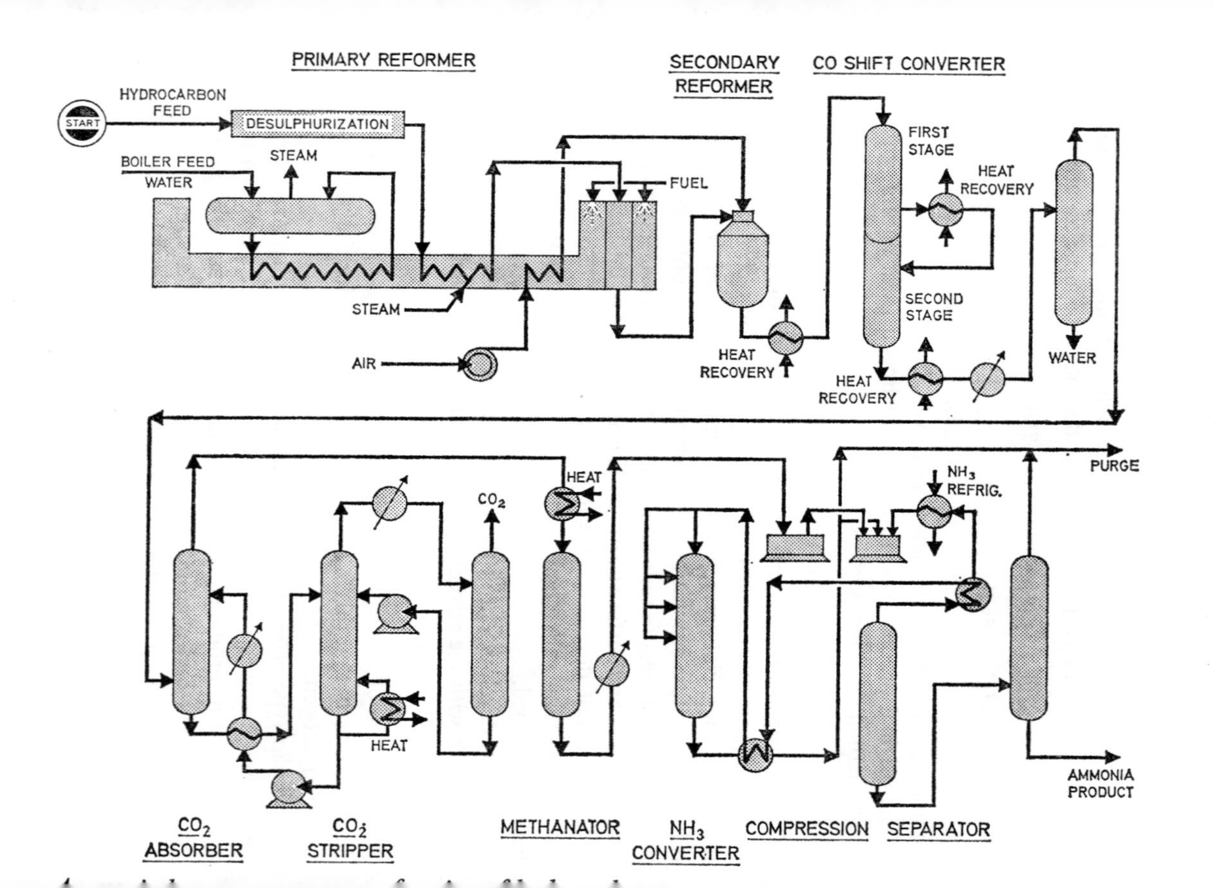
PRIMARY REFORMER
SECONDARY REFORMER
CO SHIFT CONVERTER
START
HYDROCARBON FEED
DESULPHURIZATION
BOILER FEED WATER
STEAM
FUEL
STEAM
AIR
HEAT RECOVERY
FIRST STAGE
HEAT RECOVERY
SECOND STAGE
HEAT RECOVERY
WATER
CO_2
HEAT
PURGE
NH_3 REFRIG.
HEAT
AMMONIA PRODUCT
CO_2 ABSORBER
CO_2 STRIPPER
METHANATOR
NH_3 CONVERTER
COMPRESSION
SEPARATOR

The methane–steam reaction is carried out over a nickel catalyst, so that it is necessary to eliminate sulphur compounds from the methane feed. The overall reaction, which is highly endothermic, is carried out in a tubular furnace (or primary reformer) at about 700–950 °C. It does not, of course, proceed simply in line with the single equation given. An additional reaction of some importance which takes place in the primary reformer is the shift reaction:

$$CO + H_2O \longrightarrow CO_2 + H_2$$

Primary reforming originally took place at atmospheric pressure. It has not always been taken to completion (in which case the temperature is in the lower part of the range) since the next step will commonly be that of 'secondary reforming'. This operation which is a means of introducing nitrogen into the gas stream, consists of burning a controlled quantity of air in the gas. The oxygen is consumed, the gas temperature is raised, and the gas may then be passed over a nickel catalyst to complete the reaction.

Both natural gas and refinery gases are normally available at an elevated pressure. In the operation of the traditional methane–steam reaction, it was necessary to reduce the gas supply to atmospheric pressure, and then to recompress the reaction products. This is clearly illogical and costly. Two approaches have been developed towards the elimination of this economic disadvantage: the first was to modify the methane–steam process, and the second was to develop an alternative process altogether.

Following a pioneer operation at the end of 1953, the methane–steam reaction was successively modified for operation at increasing pressure. The reforming step is now carried out at nearly 35 atmospheres and the use of pressures up to at least 50 atmospheres has been mooted. The molar ratio of steam to carbon in the feed has also been gradually reduced during the past decade from a figure of between 4 and 6 to an average of 3.

The second fundamental change has been the substitution of naphtha for methane as the hydrocarbon feed, in many parts of the world. Techniques now exist to cope with liquid feedstocks with a final boiling point of up to 220 °C. It is expected that improved catalysts will extend the range of feedstocks up to materials with a final boiling point approaching 350 °C by the mid-1970's.

In all these operations it is necessary to remove any sulphur from the feed, and this becomes particularly important with naphthas, which almost always contain sufficient sulphur to have a disastrous effect on the reforming catalyst. Traditional techniques for desulphurizing a gas feedstock include passing the gas through beds of activated carbon or zinc oxide, or wash columns using caustic soda or alkanolamines, according to the nature and extent of the sulphur compounds present. Desulphurizing a liquid feedstock may employ an acid wash technique, but it is now generally preferred to use a catalytic hydrogenation, or hydrodesulphurization, at 400 °C and 40 atmospheres. The catalyst used may typically be cobalt molybdate. Sulphur compounds are converted to hydrogen sulphide which may be recovered if this is economic. It is also possible to use alternate beds of zinc oxide and cobalt molybdate at 350–450 °C for the catalytic hydrogenation.

These changes have required a reconsideration not only of the feed preparation and reforming steps, but also of the subsequent purification process.

For operating the methane–steam reaction at 30 atmospheres and up to 950 °C highly active nickel catalysts on a refractory base are employed. Where the hydrocarbon feed is naphtha, such a catalyst would not be suitable, since carbon deposition would be excessive unless the steam to hydrocarbon ratio is raised to wholly uneconomic levels. For the pressure steam reforming of naphtha, the catalyst used is less active, containing a lower proportion of nickel, and it incorporates an alkali or alkaline earth metal promoter on a refractory base. Potassium is normally used, combined into the alumina-silica support.

To put into practice the inherent advantages of the steam reforming of hydrocarbons at elevated pressure has required an elaborate study of the engineering design of reforming furnaces, to meet the increasingly severe conditions imposed upon the furnace tubes and manifolds, and also to keep to a minimum the overall energy requirements of the operation. Alloys based on 25 per cent chromium, 18 per cent nickel steel are commonly used for the construction of furnace tubes.

With the introduction of pressure reforming, it is clear that the equilibrium of the hydrocarbon–steam reaction will be affected unfavourably, since there will always be more molecules of product than reactant. The gas stream from the primary reformers will commonly contain 6–8 per cent hydrocarbon. This places more emphasis on the secondary reforming step to which some reference has already been made.

Temperatures up to 1 000 °C are common in this operation and figures up to 1 300 °C have been noted. It is therefore appropriate to use a nickel catalyst, containing less nickel than a primary reforming catalyst, supported on a silica-free refractory base. The energy consumed in effecting the combustion of the oxygen, introduced in the air at this stage, is partly recovered by the completion of the reforming reaction by this second pass over the reaction catalyst, and also by a waste heat recovery system.

The alternative basic process, which was developed at approximately the same time as the methane–steam process under pressure, is that of partial oxidation. The reaction, expressed in terms of methane, is:

$$CH_4 + \tfrac{1}{2}O_2 \longrightarrow CO + 2H_2$$

Since dilution of the gases with nitrogen at this stage is disadvantageous in several respects, it is common to use oxygen rather than air as the basis for this reaction. This calls for an air separation unit. Steam is also commonly introduced into the reaction. Where the hydrocarbon involved is a heavy fuel oil, the hydrogen to carbon ratio is much lower than for methane (the mol ratio is rather less than 2 : 1 for a fuel oil against 4 : 1 for methane). In such a case the complete reaction of the carbon present with oxygen to form carbon monoxide will provide more than sufficient heat to raise the reaction mixture to its required temperature. In these circumstances, the addition of steam to the reactants is necessary.

In practice, as usual, a combination of reactions takes place. Important amongst these is the complete and rapid reaction of the oxygen with some of the methane to give carbon dioxide and water:

$$CH_4 + 2O_2 \longrightarrow CO_2 + 2H_2O$$

The carbon dioxide and water will react more slowly with excess methane to give mixtures of carbon monoxide and hydrogen:

$$CH_4 + CO_2 \longrightarrow 2CO + 2H_2$$
$$2CH_4 + 2H_2O \longrightarrow 2CO + 6H_2$$

The operating conditions in the combustion chambers, or gas generators, may vary appreciably. This derives at least partly from the fact that the process may be applied to almost any hydrocarbon, from methane to the heaviest fuel oils. The temperature may reach over 1 500 °C. It is possible to operate the process at atmospheric pressure, but advantage is normally taken of the fact that the reaction can be carried out under pressure, and this may range up to 130–140 atmospheres. By this means, as has already been indicated, specific advantage can be taken of the frequent availability of refinery gases and natural gas at an elevated pressure. With fuel oil as the raw material it is far more economic to pump the liquid hydrocarbon into the gas generator at an elevated pressure, rather than to effect an equivalent compression of the gaseous reaction products. The use of fuel oil as raw material will require the introduction of some means for the separation and disposal of carbon produced. The conditions of high temperature and carbon deposition do not tend to favour the use of catalysts in this operation.

The battle which was fought out between pressure steam reforming and partial oxidation in the late 1950's and early 1960's appears now to have swung somewhat to the advantage of pressure steam reforming. Where natural gas is abundantly available this is still the favoured raw material, but the advent of naphtha as a suitable and economical raw material greatly enlarged the geographical areas to which such a process could be applied. The capital cost of a pressure steam reforming plant is likely to be slightly lower than that of an equivalent partial oxidation plant (since the latter requires an air separation unit) and as now operated it is marginally more efficient, in terms of effective utilization of the total energy input. Both types of process continue in widespread operation and the balance of advantage could easily swing again. The appearance of additional supplies of natural gas in such areas as the North Sea may lend itself best to pressure steam reforming. If, however, some of the prospective additional demands for naphtha in the U.S.A. were to materialize, this could easily tilt the economic balance away from naphtha usage and towards residual fuel for ammonia production. The prospective shortage of natural gas supplies in the U.S.A. may develop new markets and new economic trends in liquid fuels. Both pressure steam reforming and partial oxidation processes have been applied to the production of town gas.

The gas streams emerging from the secondary reforming step in pressure steam reforming comprise some 50–60 per cent vol.

hydrogen, 8–15 per cent carbon monoxide, 8–15 per cent carbon dioxide, 20–25 per cent nitrogen, and traces of oxygen and hydrocarbon. From partial oxidation the product gas streams typically contain 45–60 per cent vol. hydrogen, 30–47 per cent carbon monoxide, 2–5 per cent carbon dioxide, together with traces of nitrogen and hydrocarbon. In the latter case there can also be traces of sulphur compounds which must be removed.

The first processing step applied to either of these gas streams is the shift reaction. This has been modified out of all recognition in the past few years. The conventional shift conversion was carried out at 350–500 °C over an iron catalyst promoted with chromium, and this reduced carbon monoxide content to about 3 per cent. More recently the conversion has been carried out at elevated pressure (e.g. 30–35 atmospheres) to tie in with the generation of the gas streams at elevated pressure. Increasing pressure improves catalyst activity and so does increasing temperature, but this also decreases the equilibrium constant. An intermediate stage in this development was a two-stage shift conversion with the conventional catalyst and this could reduce the carbon monoxide content of the stream to about 0.5 per cent.

The new low temperature conversion catalysts have enabled a more efficient and simple system to be employed. They are generally based upon copper, zinc and chromium in combination. They operate at 200–250 °C.

The present procedure for shift conversion employs one stage at two temperature levels. The bulk of the carbon monoxide is converted in an adiabatic converter over an iron oxide–chromium catalyst operating at temperatures between 370 °C inlet and 430 °C exit. Pressures may be up to 27–28 atmospheres. This stream then passes over a copper–zinc–chromium catalyst at the same pressure and at temperatures in the range of 200–250 °C. There is a temperature rise of 15 °C in this operation. The carbon monoxide content of the stream is reduced to 0.2 to 0.3 per cent.

The low temperature catalyst is prepared as a mixture of oxides, and to make it active the copper oxide has to be reduced to metallic copper. This is a highly exothermic reaction calling for much care. The catalyst so produced is sensitive to sulphur and chlorine at concentrations as low as 1 to 2 parts per million. Special provision, in the form of a zinc oxide guard chamber, is included in most designs to obviate the danger of the carry over of traces of sulphur from the high temperature conversion catalyst during start up. The catalyst is also sensitive to temperature. During the reduction step the highest allowable temperature is 220–230 °C. In general operation a temperature of 260 °C or above will deactivate the catalyst. When the catalyst has been activated it is pyrophoric and if the converter is for any reason to be exposed to air, the catalyst must be oxidized with an inert gas containing a small proportion of oxygen. Here again there is a temperature limit of 220–230 °C, if the catalyst is to be capable of being activated again.

After shift conversion, the only major remaining impurity in the gas stream is carbon dioxide. The early method of carbon dioxide removal was by scrubbing with water at pressures of 15–20 atmospheres.

The water-scrubbing technique was replaced a number of years ago by an alkanolamine absorption. This usually employed monoethanolamine which could reduce the carbon dioxide content of the stream to 50 parts per million, whereas water scrubbing could only remove carbon dioxide down to 0.5 per cent of the stream. A further development has been the Sulfinol process in which the absorption medium is likely to be a combination of diisopropanolamine and sulfolane with water.

The generally preferred processes today for this operation are modifications of the original Benfield process, using hot potassium carbonate solution as the means of absorption. Among the reasons for this move have been the fact that the heat required for regeneration of the solvent is about half that required for monoethanolamine, the capital cost of the plant is less, and the treatment of gas entering the absorber at 110 °C saturated with water vapour makes these processes easier to fit into an integrated energy-balance flowsheet.

The absorbent is a 28–30 per cent potassium carbonate solution at 100–120 °C. Such processes may be operated in two stages in which the process gas is finally scrubbed with a portion of the circulating liquor which has been regenerated more completely than the bulk of the liquor. By this means the proportion of carbon dioxide in the product gas is kept down to 0.1 per cent.

The most widely used process of this type is the Giammarco–Vetrocoke process which incorporates arsenic oxide to increase the rate of absorption and desorption. It also assists in the control of corrosion. The presence of arsenic causes the liquor to be highly toxic and special precautions to prevent leakage are required. Other additives are recommended in various process modifications, particularly to minimize corrosion.

It was traditional to effect the final removal of carbon monoxide by absorption in 'copper liquor' (a solution of a basic copper salt such as copper ammonium acetate). This is a high pressure operation which is expensive in utility consumption and involves some loss of hydrogen and nitrogen.

With the development of low temperature shift conversion, which reduces carbon monoxide content to a relatively low level, it is possible to obviate an absorption step, and revert to a modification of the long established process of methanation. This is a very simple process, passing the gas stream containing carbon monoxide over a nickel catalyst at 270–400 °C. The carbon monoxide reacts with hydrogen present to form methane and water. Traces of carbon dioxide react similarly. Since this operation consumes hydrogen, and forms an inert gas which will ultimately require to be purged from the ammonia synthesis loop, it is clear that it can only be applied to gas streams with the content of oxides of carbon already reduced to a low level. Originally methanation was used in ammonia processes using a high synthesis pressure, since the methanation step itself required a high pressure. Catalyst improvements have now made it possible to carry out this operation at pressures as low as 21 atmospheres.

In plants carrying out the partial oxidation process it is necessary to incorporate an air separation unit to provide oxygen. Advantage can be taken of this to introduce nitrogen into the synthesis gas by the rather elegant nitrogen wash technique. In this type of operation, nitrogen is introduced after the hydrogen stream has been largely purified. The nitrogen is fed to a liquid nitrogen wash column where traces of carbon monoxide, argon and methane are removed in solution in liquid nitrogen from the base of the column. The purified hydrogen stream, together with a proportion of evaporated nitrogen, passes out at the top. Further nitrogen is added to the stream to make up to the appropriate proportion for ammonia synthesis. The removal of inert gases by the nitrogen wash is an advantage in the synthesis step, in that the need for a purge (with consequent loss of useful material) is reduced to a minimum.

The actual ammonia synthesis is carried out in a very complex conversion, refrigeration and recycle system. The catalyst is essentially iron normally promoted with oxides of aluminium and potassium. Conditions have varied widely, but until recently some degree of uniformity had been reached by many processes, using pressures of 270–340 atmospheres and temperatures of about 500 °C. This gives 20–25 per cent conversion to ammonia per pass.

The design of huge single train ammonia plants with capacities of around 1 500 tons ammonia per day has given rise to significant new developments. Centrifugal compressors are less costly items of equipment than reciprocating compressors, especially in the very largest sizes, where one centrifugal may replace multiple reciprocating compressors. Centrifugals may also incur a lower maintenance charge, but in the smaller sizes they are less efficient in operation and are larger users of utilities than reciprocating compressors. They can lend themselves to an overall plant integration of energy, in that they can be powered by steam turbines driven by steam from waste heat boilers.

In order to take advantage of the economic benefits which centrifugal compressors could provide, a stage was reached where synthesis pressures were deliberately reduced to about 150 atmospheres, even though this would clearly reduce the efficiency and increase the cost of the synthesis step.

Improvements in centrifugal compressor design have removed most of these constraints. Such compressors are capable of coping with single-train plants of 1 500 tons per day capacity, and discharge pressures up to 550 atmospheres are feasible.

Current synthesis converters are once again operating in the temperature range of 450–500 °C and at pressures of 270–350 atmospheres.

In the current design of synthetic ammonia plants there are a number of characteristic features—a high capacity single-train operation; high pressure reforming; centrifugal compressors; all major power requirements served by steam turbine; minimal electricity consumption. The aim of this exercise is to develop an integrated heat recovery and energy system, whereby most of the power and heating requirements are met by heat recovery from high temperature gases at the reformer stage.

A source of continuing interest for ammonia production is the volatile fraction of the gas stream from catalytic reforming units. There is commonly available from a catalytic reformer a fraction comprising 80–95 per cent volume of hydrogen, with methane as the major impurity. Such a stream, if available in sufficient quantity, is clearly valuable for ammonia production. The introduction of nitrogen is a factor which may vary according to specific conditions. It may be convenient to consume the hydrocarbons present in a process akin to secondary reforming, and to introduce nitrogen by this means. If a liquid nitrogen wash is to be employed, it is only likely to prove economic if an application exists for the oxygen from the air separation unit. Methane will be separated from the hydrogen stream in the liquid nitrogen column. The methane may then form part of an additional synthesis gas train operating a partial oxidation process, consuming oxygen. Whatever detailed arrangements are made to suit local conditions, the size and complexity of the purification units are likely to be less with such a concentrated hydrogen stream than with conventional synthesis gas preparation.

The production and consumption of ammonia is traditionally expressed in terms of metric tons of nitrogen content. World statistics for nitrogen (in '000 metric tons N) are as in the following table. The fertilizer year extends from July 1st to June 30th.

	Production	*Total consumption*	*Production for industrial use*
1955/56	8 068	7 552	1 243
1959/60	11 057	10 459	1 715
1965/66	23 188	22 452	3 437
1969/70	40 200	36 910	7 320
1974/75 (est.)	55 130	53 990	9 970

It will be appreciated that these figures include modest quantities of naturally occurring nitrates, of calcium cyanamide, and of by-product ammonium sulphate, but the great preponderance of the totals reflects the vast and growing importance of synthetic ammonia. The proportion of the total nitrogen usage finding an industrial application has not changed very much. It has crept up from about 16 per cent to nearly 20 per cent over the past decade, but seems likely to remain near this figure. There may be some confusion in the figures relating to the production, for example, of caprolactam or acrylonitrile. Here the ammonia undoubtedly finds an industrial outlet but some of it reaches the fertilizer market ultimately as ammonium sulphate by-product.

Before 1939 over 90 per cent of world synthetic ammonia capacity was based on coal or coke. This had dropped to 40 per cent in the 1958–59 season, by which time petroleum accounted for 55 per cent of the total. It was recently estimated that by 1971–72 the proportions of world ammonia capacity would be:

Coal	8.7%
Naphtha and other petroleum liquids	27.2%
Natural gas	56.1%
Others	8.0%

The main element in 'others' is likely to arise from electrolytic hydrogen.

For a considerable number of years a very high proportion of U.S. synthetic ammonia has been petroleum-based. The first major plant to use a petroleum feedstock for synthetic ammonia production in the United Kingdom came on stream as recently as 1957, but virtually all ammonia production in this country had switched to a petroleum basis by the middle 1960's. A significant proportion of this production is now based on North Sea natural gas.

While the fertilizer market, as is common in the field of agricultural commodities, has its ups and downs, there seems to be no break in the continuous pattern of growth in the usage of synthetic ammonia. Two factors have stimulated this growth; one, a continuing factor, is the urgent need to expand the world's total output of food, particularly in the developing countries, and the second is the significant reduction in the cost of producing ammonia obtained in the past two decades, in the face of a continuous period of cost inflation. The development of the petroleum-based processes has played a part in this. Petroleum raw materials have a natural advantage in being fluids, and in providing a raw synthesis gas rich in hydrogen. The range of processes is such that an appropriate raw material can be made available economically anywhere it may be required. These natural advantages have been supplemented by a triumph of continuous technical innovation in a basically well-established process, and by the successful exploitation of the benefits of scale. It will be difficult to hold back the waves of inflation for much longer.

Ammonia represents the largest scale of production of any chemical made from petroleum.

The techniques indicated (more particularly the methane–steam reaction) represent a means of preparing hydrogen streams for all types of hydrogenation operation, but by far the largest volume of hydrogen is required for ammonia production.

Incidental to the question of the source of raw material for ammonia, the whole fertilizer industry has been consolidating a number of trends during this recent period of development. One factor of interest is the direct injection of ammonia or its solutions into the soil and into irrigation water. This is an effective way of fixing nitrogen in the soil without the necessity for making further derivatives from ammonia. These techniques are now widely applied in the U.S.A. and owe their commercial development to petroleum chemical ammonia.

The trend in fertilizers is towards mixtures of high plant nutrient content. This has led, in production terms, to a concentration on such products as ammonium nitrate and urea, and away from ammonium sulphate.

Both ammonium nitrate and urea production units are normally adjuncts to ammonia plants. Ammonia is readily oxidized to nitric acid, which is then reacted with more ammonia to give ammonium nitrate.

To oxidize ammonia to nitric acid, a gas mixture of 10 per cent ammonia with air is passed through platinum or platinum-rhodium gauze at high temperature. The essential product of this reaction is nitric oxide. The next stage is an oxidation and absorption to produce nitric acid.

$$4NH_3 + 5O_2 \rightarrow 4NO + 6H_2O$$

$$2NO + O_2 \rightarrow 2NO_2$$

$$3NO_2 + H_2O \rightarrow 2HNO_3 + NO$$

Earlier processes, particularly in Europe, carried out both these steps at atmospheric pressure. Processes developed in the U.S.A. tended to carry out both steps at a pressure of around 8 atmospheres. A third type of process represented a compromise, whereby the oxidation was carried out at atmospheric pressure and the absorption at an elevated pressure.

Atmospheric oxidation has the advantage of an improved yield of acid on the ammonia, and considerably lower loss of platinum catalyst. The pressure operation reduces the size (or, in the case of multiple units, the number) of the equipment items, so that capital cost is reduced. The final concentration of acid is also slightly higher.

The oxidation is carried out at a temperature of 750–800 °C in the case of atmospheric oxidation. For pressure oxidation the temperature used is around 900 °C and multiple thicknesses of gauze are employed.

Pressure absorption is markedly more efficient than the process carried out at atmospheric pressure. The combination of atmospheric oxidation and pressure absorption involves a heavy cost for a stainless steel compressor to handle the aqueous nitrogen oxides.

As with so many other processes, careful attention must be paid to the energy balance. The ammonia oxidation, the nitric oxide oxidation, and the reaction to produce nitric acid all generate heat, and maximum recovery of this energy is important. With pressure absorption it is also possible to recover power from the exhaust gases by gas turbine or similar means.

With atmospheric absorption the concentration of acid produced approximates to 50 per cent. With pressure absorption a concentration of 57–60 per cent is possible.

To obtain higher concentrations of nitric acid it is first possible to reach a concentration of 68 per cent acid by distillation. This produces a water azeotrope at the 68 per cent acid level. It is possible to concentrate the acid further by distillation alternately at atmospheric and elevated pressure, since the composition of the azeotrope varies with pressure, but it is generally more economic to achieve concentration of nitric acid beyond the 68 per cent level by means of a dehydrating agent. Sulphuric acid at 93–98 per cent is normally used as the dehydrating agent.

Nitric acid as produced in a pressure absorption unit (i.e. at 57–60 per cent concentration) may be reacted directly with gaseous ammonia. This generates a considerable amount of heat which may be used to concentrate the ammonium nitrate solution produced to around 83 per cent. Further concentration may be achieved by evaporation or spray drying. The process for spraying moist molten salt to produce dry granules for fertilizer use is generally described as 'prilling'. In view of the explosion hazard involved in the handling of ammonium nitrate, the solid granules are normally coated with an inert material, or the salt mixed with lime or limestone to produce a blended fertilizer.

Ammonia plants involve the separation of a carbon dioxide stream from the synthesis gas, as previously described. Frequently this is simply exhausted to atmosphere. Alternatively, carbon dioxide may be compressed for marketing in the liquid or solid form. The most satisfactory outlet for this stream, however, is in its reaction with ammonia to produce urea.

The production of urea proceeds in two steps, the first being the reaction of excess ammonia with carbon dioxide (conventionally at 160–220 °C and 180–350 atm.) to form ammonium carbamate. This is an exothermic reaction. In the same reactor the ammonium carbamate dehydrates to urea to the extent of around 60–70 per cent. The numerous processes differ in their technique for the separation and recycling of the unreacted ammonium carbamate or its constituents.

Mention should be made of the recently developed 'low' pressure process (130–150 atm.) of DSM. The 'stripping' of the reactor effluent with fresh carbon dioxide at synthesis pressure resolves the problem of subsequent handling of ammonium carbamate by decomposing nearly all of it into its constituents, ammonia and carbon dioxide. The two gases, together with fresh ammonia, are then recycled to the reactor.

$$2NH_3 + CO_2 \longrightarrow \underset{\text{ammonium carbamate}}{NH_4CO_2NH_2}$$

$$NH_4CO_2NH_2 \rightleftharpoons \underset{\text{urea}}{NH_2CONH_2} + H_2O$$

The process of SNAM Progetti also uses a relatively low pressure (150 atmospheres) in the reactor. This process employs a 'stripper', using ammonia as the stripping medium, in order to decompose unreacted ammonium carbamate in the reactor effluent. The Japanese (Mitsui Toatsu) process employs two decomposers in series, the first at high pressure and the second at low pressure, to achieve the same objective.

The other critical feature in urea production, especially for material to be used as fertilizer, is to minimize the formation of biuret (a condensation product formed from two molecules of urea). Efforts are made to avoid conditions favourable to biuret formation during the evaporation stage, and these may be supplemented by a vacuum crystallization step to separate biuret after it has been formed.

Urea has attained a big tonnage and in the U.S.A. alone about 3.5 million tons were consumed in 1970. To this total imports contributed sustantially. Urea has the advantage of its high nitrogen content, and also its high solubility in water, which enables it to be applied to soil as a solution. Some 37 per cent of U.S. usage in 1970 was for liquid fertilizer, and 35 per cent as a solid fertilizer. A further 15 per cent was used as a supplement to cattle food. The remaining 13 per cent finds largely non-agricultural uses—notably in urea-formaldehyde and urea-melamine resins.

Melamine is traditionally a coal-based chemical, being derived ultimately from calcium carbide. Much attention is now being paid to producing it from urea. Earlier attempts using high pressure processes were not particularly successful economically, but more is expected of the low pressure processes now developed. A process operating essentially at atmospheric pressure involves two steps:

$$(NH_2)_2CO \longrightarrow HNCO + NH_3$$

$$6HNCO \longrightarrow \underset{\text{melamine}}{C_3N_3(NH_2)_3} + 3CO_2$$

The first step is a decomposition of urea at 350 °C. The urea in solid form is conveyed to the decomposer in a flow of ammonia gas. The decomposition to cyanic acid and ammonia is non-catalytic. The solid urea decomposes almost immediately to gaseous cyanic acid which is passed over a fixed bed of catalyst at 400–450 °C. The temperature is high enough to sublime the melamine produced. The stream is quenched by direct contact with water and a melamine suspension is formed. Melamine is separated by centrifuging, and dried. Another process uses 5–10 atm. pressure and feeds molten urea into the single stage reaction where it contacts the fluid bed catalyst. Here again ammonia acts as the carrier gas and fluidizing agent.

A further recent contender in this field is the B.A.S.F. process. This is said to operate at atmospheric pressure and 380 °C in the presence of fluidized alumina as catalyst. The urea, which is in the vaporized form, is converted to the extent of 95 per cent. The reactor effluent gases are cooled to a point above the temperature at which melamine will sublime, and impurities which solidify are removed by hot filtration. The gases then pass to a sublimer, held at 170–200 °C, where 98 per cent of the melamine crystallizes. Unreacted urea is recycled. There is clearly much merit in the one-step reaction from urea to melamine carried out at atmospheric pressure.

The production of melamine in the U.S.A. in 1970 was about 34 000 long tons. There was still more capacity in the U.S.A. at the end of 1970 for melamine from dicyanamide (i.e. from calcium carbide) than from urea. All new capacity is urea-based and dicyanamide plants are expected to shut down as this capacity builds up. The U.S. melamine production is supplemented by significant imports, so that the consumption of melamine there in 1970 could have reached over 45 000 long tons. Of this total some 37 per cent was used for high pressure laminates, 24 per cent for moulding compounds, 12 per cent for surface coatings, 18 per cent for textile and paper treating resins and the remaining 9 per cent found miscellaneous uses.

The whole field of ammonia derivatives is a wide and complex one. It is not proposed to pursue the subject beyond the brief outlines given above.

Methyl Alcohol

The distillation of wood was the original source of methyl alcohol. This process has been largely, but not entirely, eliminated.

The switch to coal as the main source of methyl alcohol was initiated by the Germans, and attained a major commercial significance in the 1920's.

More recently, the trend towards the use of petroleum sources of raw material for synthesis gas has followed a very similar pattern for both ammonia and methyl alcohol. Today almost all the world's methyl alcohol is petroleum-derived.

A minor but significant proportion of methyl alcohol of petroleum origin in the U.S.A. is obtained by the oxidation of higher paraffins. The major developments have been based upon natural gas, refinery light gas streams, or naphtha fractions.

The basic reactions for the synthesis of methyl alcohol are as follows:

$$CO + 2H_2 \rightarrow CH_3OH$$
$$CO_2 + 3H_2 \rightarrow CH_3OH + H_2O$$

The crude synthesis gas is treated somewhat differently in this

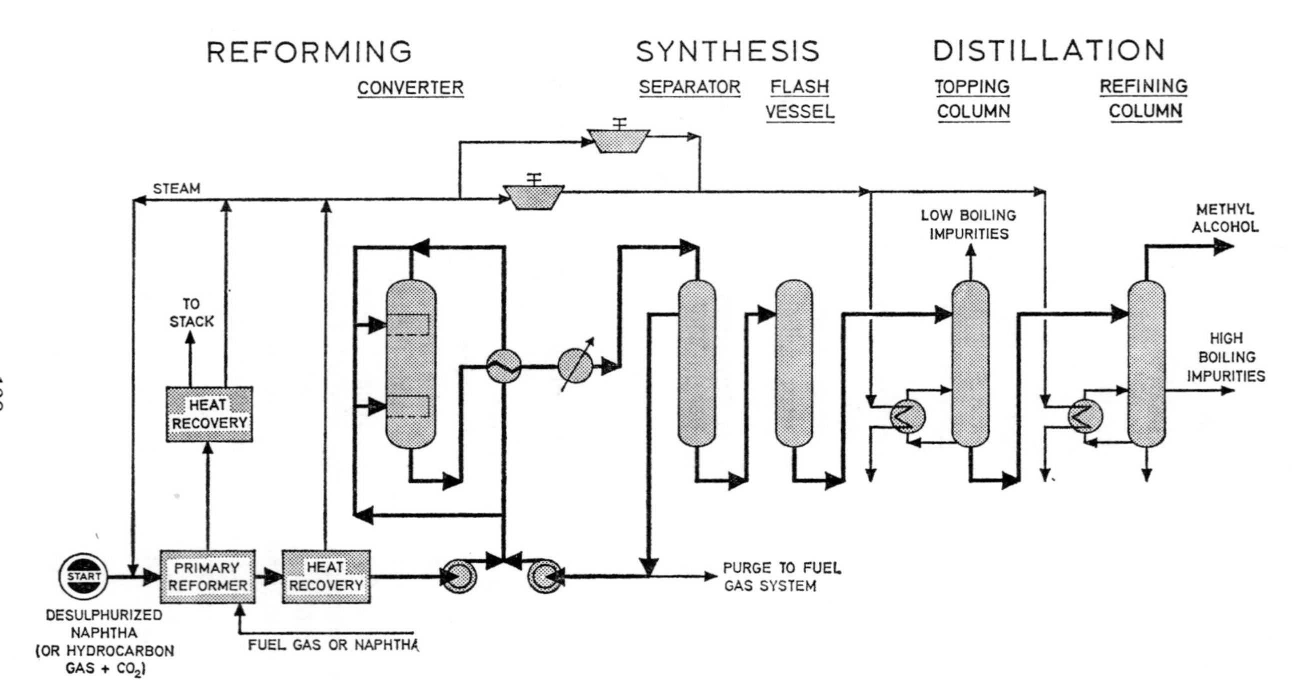

Methyl alcohol by low pressure process

(From *Hydrocarbon Processing*, Nov. 1971)

case than in ammonia synthesis. Carbon monoxide is one of the synthesis raw materials, and therefore the shift reaction is not applied. There is, of course, no requirement in the case of methyl alcohol synthesis for the introduction of nitrogen.

In the manufacture of crude synthesis gas the proportions of hydrogen and carbon monoxide can be varied by an adjustment of the conditions of gas generation. In the methane–steam reaction, for example, the use of excess steam will push the equilibrium in favour of carbon monoxide. A more marked effect can be obtained by the recycle of a carbon dioxide gas stream to the synthesis gas generation stage. Such a carbon dioxide stream may be obtained, for example, from an adjacent ammonia plant, or alternatively it may be extracted, using the normal absorbents, from the flue gases arising from the reformer furnaces. By such means the conditions for the manufacture of the crude synthesis gas are adjusted to give an appropriate relationship between the proportions of hydrogen and oxides of carbon present, as indicated by the synthesis reactions.

At one time it was common to rely essentially on the reaction between hydrogen and carbon monoxide, even to the extent of removing carbon dioxide from the reformed gas stream. Now both oxides of carbon are utilized in the overall reaction.

In the conventional process the mixed gas stream is compressed to 275–360 atm. pressure and heated. The gases then pass to a complex converter similar in essence to those used for ammonia synthesis. In the tubes of the 'catalyst basket' the conventional catalyst is based on zinc oxide, normally promoted with other metal oxides, of which chromium oxide is the most significant. The reaction temperature is controlled within the range 300–400 °C by the provision of appropriate facilities for heat removal.

The equilibrium reaction is affected favourably by elevated pressures, and adversely by an increase of temperature. The conventional catalysts will operate effectively within the desired temperature range only when the reaction proceeds at high pressures.

Methyl alcohol is condensed out of the reaction products, and the unreacted gases recycled. For the production of the synthesis gas stream, the pressure steam reforming of natural gas or naphtha is normally preferred. Where sulphur is present in the feedstock (as is likely with naphtha) a desulphurization step must be applied before it is admitted to the reformers.

As was mentioned in connection with developments in the ammonia process, there is considerable merit in being able to use centrifugal compressors for these large-scale high pressure plants. The difficulty with methyl alcohol was that the efficient use of centrifugal compressors, at the pressure level employed in the conventional process, was only feasible for very large plants. This is also the case with ammonia, but ammonia is produced on a much greater scale than methyl alcohol, so that there was room for a significant number of huge new ammonia units. With methyl alcohol the scope for building vast new plants was much more limited.

It was most timely, therefore, that just as this problem was causing concern, I.C.I. came forward with their low pressure process for methyl alcohol. This involves a catalyst understood to be based on copper, zinc and aluminium. The normal operation of the process uses a converter pressure of 50–60 atmospheres. A somewhat lower temperature of reaction is used than in the conventional process—about 240–260 °C. Another feature of the I.C.I. process is a modified converter, which enables the process of catalyst charging and discharging to be much simplified and speeded.

There is nothing basically new about the use of a copper-based catalyst for this reaction. It was known that such catalysts were effective, and indeed, a Japanese company claims to have used a copper-based catalyst for over 30 years. Generally, however, the copper catalysts were regarded as too easily poisoned and too readily deactivated by heat to be a commercial proposition. Once the way was pointed, it was to be expected, therefore, that there would be additional processes entering the low-pressure methyl alcohol arena. The alternative processes all make use of a copper-based catalyst. The Nissui–Topsoe process uses a copper–zinc–chromium catalyst at 230–260 °C and 100–150 atmospheres pressure. Lurgi proposes a copper-based catalyst at 40 to 60 atmospheres. Catalyst and Chemicals Inc. put forward 240–250 °C and 100–250 atmospheres over a copper–zinc catalyst as their recommended basis. It will be noted that some of these processes can scarcely justify the description ‘low pressure’, but are generally designated ‘intermediate pressure’. I.C.I. themselves have put forward a variant of their basic process which operates at 100 atmospheres and uses a modified catalyst.

The vital feature of all these new processes is that a plant of moderate size by today’s standards can operate with a centrifugal synthesis gas compressor. The conventional process seems to be still competitive so long as the plant required is a very large one. The intermediate pressure plants can claim some advantage in reduced equipment size for large-scale units compared with the low pressure processes, which will, of course, always have some advantage in reduced energy consumption.

Although it is no longer common to integrate ammonia and methyl alcohol production to any great extent, there are sites where a range of processes using synthesis gas, or possibly hydrogen and carbon monoxide separately, may be concentrated. There are some additional processes (notably partial oxidation processes for acetylene production) where the 'off-gas' may be utilized in such a complex.

Methyl alcohol has been one of the more commercially volatile products among the petroleum chemicals. It switched from shortage to generous supply more than once in the later 1960's and early 1970's. The latter half of 1970 saw a lull in demand just as the development of the low-pressure processes was stimulating an interest in new building. A considerable volume of new capacity was scheduled for completion in the years 1971 and 1972. In total this could amount to some 3 million long tons per year in the U.S.A. and western Europe, raising the total nominal capacity in these combined areas to about 8 million tons per year. This was naturally offset by some plant redundancies. Nevertheless, it was freely forecast that over-capacity for methyl alcohol would last until 1974 at least. It was, therefore, interesting to observe that a surge of increased demand in the U.S.A. during 1972, coupled with some plant difficulties, created at least a temporary shortage in mid-1972.

The most detailed recent breakdown of methyl alcohol demand comes from Japan:

Formaldehyde	53%
Dimethyl terephthalate	8%
Methyl methacrylate	6%
Synthetic resins	5%
Pharmaceuticals and agricultural uses	4%
Surface coatings	3%
Methylamines	3%
Polyvinyl alcohol	2.5%
Chloromethanes	1.5%
Miscellaneous	14%

The Japanese consumption of methyl alcohol was nearing 1 million long tons in 1970 and rose above this figure in 1971.

A corresponding picture for the U.S.A. for 1969 with a forecast for 1975 is as follows:

	1969	*1975* (est.)
Formaldehyde	47%	49%
Dimethyl terephthalate	9%	11%
Solvents	7%	7%
Methylamines	4%	3%
Inhibitor	4%	4%
Methyl methacrylate	3%	4%
Methyl halides	4%	5%
Miscellaneous	22%	17%

The U.S. production of methyl alcohol in 1970 was 2.04 million long tons. In the United Kingdom the production was around 360 000 long tons and in France about 210 000 tons.

Overall the percentage of methyl alcohol used to make formaldehyde is just about 50 per cent, and this figure remains fairly constant in all major industrial countries.

The usage of formaldehyde in the U.S.A., where it is expressed in terms of the normal 37 per cent product, was about 1.95 million long tons in 1971. The breakdown of applications in the U.S.A. in 1971 approximated to the following:

Phenolic resins	25%
Urea–formaldehyde resins	25%
Melamine resins	8%
Hexamethylene tetramine	6%
Polyacetals (polyformaldehyde)	8%
Pentaerythritol	7%
Fertilizers	5%
Acetylenic chemicals	2%
Miscellaneous including exports	14%

Formaldehyde lost a market (which amounted to some 135 000 tons per year in the U.S.A. in the mid-1960's) when du Pont shut down their plant for making ethylene glycol by way of glycollic acid (see Chapter 7) in 1968. Another vulnerable market for formaldehyde is for hexamethylene tetramine. The use of this material, which is almost entirely for explosives, was greatly boosted by the Vietnam war, and production, already showing a reduction from the 1969 figure, will drop further once peace is established. Most other outlets for formaldehyde are growing steadily—about half the total usage is for some form of building construction material.

About 16 per cent of the U.S. formaldehyde capacity is by way of paraffin oxidation. More generally, at least 90 per cent of the world's formaldehyde is made from methyl alcohol.

The United Kingdom production of formaldehyde, expressed modestly in terms of 100 per cent material, was 121 700 long tons in 1970.

Formaldehyde is normally made by the catalytic oxidation/dehydrogenation of methyl alcohol. The catalyst commonly comprises silver or copper on an inert support, and the temperature is in the range 450–600 °C. It is customary to employ a slight deficiency of air for the completion of the oxidation reaction.

$$CH_3OH + \tfrac{1}{2}O_2 \rightarrow CH_2O + H_2O$$

This causes some formaldehyde to be produced also by dehydrogenation.

$$CH_3OH \rightarrow CH_2O + H_2$$

The second of these reactions is endothermic, whereas the first is highly exothermic. Adjusting the conditions so that both reactions take place enables the net reaction to become thermally self-supporting.

An alternative process, which involves essentially the oxidation reaction of methyl alcohol to formaldehyde, requires an excess of air in the presence of a catalyst comprising a mixture of metal oxides, commonly molybdenum oxide promoted with iron, at 350–450 °C. The yield of formaldehyde on methyl alcohol fed, at around 90 per cent or slightly more, is comparable with that of the process using a metal catalyst. As the temperature of the reaction is lower and the catalyst may be cheaper, there is some economic incentive to consider the metal oxide type of catalyst.

As the oxidation reaction to produce formaldehyde is strongly exothermic, it is normal to generate steam or otherwise recover useful heat from the reaction products.

The oxide catalysts are gaining some ground in this field but still represent only a minority of the total formaldehyde units.

A further methyl alcohol reaction which requires comment is that with carbon monoxide to produce acetic acid. This has been the subject of study for many years.

$$CH_3OH + CO \rightarrow CH_3COOH$$

According to the catalyst used the reaction can be made to produce methyl formate, methyl acetate or acetic acid. Alkaline catalysts favour the production of methyl formate and acidic catalysts favour either acetic acid or methyl acetate. Earlier work concentrated on phosphoric acid catalysts promoted with copper. The development of a commercial process was hindered by the severe conditions necessary—a high pressure and temperature coupled with a considerable corrosion stress.

The first commercial process was developed by B.A.S.F. in Germany, and is another illustration of the versatility of Reppe chemistry. The conversion takes place at 250 °C and 650 atmospheres. The catalyst has been stated to be cobalt acetate and cobalt iodide in combination. The employment of corrosion-resisting alloys is a major factor in making this process workable. By-products are said to arise only to the extent of 4 per cent of acetic acid made, and consist of propionic acid, ethyl acetate, butyl acetate and smaller amounts of other products.

A modification of this process has been developed by Monsanto in the U.S.A. The specific aim was to carry out the process using less severe reaction conditions. This is achieved by using an iodide-promoted rhodium catalyst. The reaction conditions are said to be up to 15 atmospheres pressure and a temperature between 175 and 245 °C. A catalyst complex is generated, which incorporates not only the rhodium and iodine, but also carbon monoxide and methyl groups. The complex breaks down on hydrolysis to provide acetic acid.

Both these processes are now in commercial operation.

Oxo Process

Reference has already been made to the significance of the Oxo process in the production of C_4 and C_8 alcohols from propylene.

The process, in its conventional form, is a means of converting an olefin to aldehydes containing one more carbon atom, by carbonylation or reaction with an approximately equimolar mixture of hydrogen and carbon monoxide. The catalyst is cobalt-based and may conveniently be added in the form of cobalt naphthenate, since this is readily available and soluble in organic media.

The reaction takes place in the liquid phase at 130–175 °C and 250 atm. Where propylene is to be reacted it may be dissolved in an appropriate high boiling inert medium. The crude aldehyde product is treated with steam at high temperature to facilitate catalyst removal.

Where an alcohol is the ultimate requirement, as it usually is, the aldehyde may be hydrogenated in the presence of a nickel catalyst at 150 °C and up to 100 atm.

One of the major applications of the Oxo process involves propylene as feedstock. It has already been pointed out (in Chapter 8) that this forms both n- and isobutyraldehyde.

$$\underset{\text{propylene}}{CH_3CH{=}CH_2} + CO + H_2 \rightarrow \underset{\text{n-butyraldehyde}}{CH_3CH_2CH_2CHO}$$

$$\rightarrow \underset{\text{isobutyraldehyde}}{CH_3CH(CH_3)CHO}$$

Originally the maximum proportion of the n-isomer was believed to be fixed at 60 per cent. It is desirable to carry out the reaction using conditions as mild as possible since low temperatures favour the formation of the n-isomer. It is also possible to recycle the branched chain isomer which tends to minimize further production according to the principle of Le Chatelier. The use of selective solvents is also claimed to favour the formation of the n-isomer. Such process variations can only be applied to a limited extent as they tend to diminish the overall plant productivity. It is generally claimed today that commercial Oxo processes will produce n- and isobutyraldehyde in the ratio of about 4 to 1.

It is possible to separate the n-butyraldehyde and subject it to an aldol condensation (similar in principle to that applied to acetaldehyde in n-butyl alcohol manufacture) in the presence of an alkaline catalyst under mild conditions. The C_8 aldehyde formed is hydrogenated in the normal manner to 2-ethylhexanol.

The Aldox process allows the dimerization to proceed without the separation of the C_4 aldehydes. The basic Oxo reaction proceeds at 200 atm. and 180 °C, using a catalyst system in which the usual cobalt salt is employed in conjunction with a zinc compound, such as zinc acetylacetonate. The dimerization (which is incomplete) may be encouraged to proceed further by a subsequent 'soaking' step, at lower pressure and at 90–230 °C, using more zinc catalyst. The aldehydes may be hydrogenated in the usual way to provide both

C_4 and C_8 alcohols. This form of dimerization is of more general application, but its use in providing C_4 and C_8 alcohols in the same basic operation is the most important practical instance.

Another new development of considerable potential importance is the low-pressure single stage Oxo process for alcohol production. This uses a lower pressure (about 30 atm.) and the catalyst required is the tributyl phosphine complex of cobalt carbonyl. Not surprisingly, as the alcohol is the desired product, the synthesis gas used should contain more hydrogen than usual. The molar ratio of hydrogen to carbon monoxide is normally 2:1. This process is claimed to produce an exceptionally high ratio of n- to isobutyl alcohol when applied to propylene. Where straight chain higher olefins in the C_{12}–C_{18} range are the feedstock, the products are straight-chain primary alcohols of interest in detergent products. The detergents obtained from such materials are readily biodegradable. This process requires larger quantities of catalyst to be present than the conventional high pressure processes, so that the economics of the catalyst recycle system are important.

An alternative catalyst complex for low-pressure Oxo reaction could be based on rhodium rather than cobalt. This can be made to minimize by-product formation, but is not effective in improving the ratio of n- to branched chain isomer. This operation has not been developed to the commercial stage.

World capacity for Oxo chemicals has continued to increase rapidly. In 1971 nearly 700 000 tons of Oxo chemicals were consumed in the U.S.A. About 50 per cent of this material was based on propylene (to produce butyl alcohols and 2-ethylhexanol), 28 per cent used C_6, C_9 or C_{12} olefins (to make plasticizer alcohols), 5 per cent was derived from C_{11}–C_{13} linear olefins, 4 per cent from ethylene (for propionaldehyde and derivatives), and 13 per cent used various other raw materials.

In the United Kingdom the long-established I.C.I. capacity at Billingham is now supplemented by a complex using the low pressure process operated by Shell at Carrington.

In the U.S.A. about 70 per cent of Oxo products end up as vinyl plasticizers, and a similarly high proportion is likely to be directed to this outlet in other countries.

Carbon Monoxide Reactions

Reference has already been made to the use of carbon monoxide as a raw material in synthesis gas, for acrylic acid from acetylene (Chapter 4), in ethylene glycol manufacture (Chapter 7), in propionic acid manufacture (Chapter 7), in the Reppe process for n-butanol (Chapter 8), and, earlier in this chapter, for acetic acid production from methyl alcohol.

Phosgene is an important material, simply made by passing pure dry carbon monoxide with chlorine over activated charcoal at 200–255 °C.

$$CO + Cl_2 \rightarrow \underset{\text{phosgene}}{COCl_2}$$

The most important factor in this operation is the provision of adequate facilities for the dissipation of the heat arising from the exothermic reaction.

The production of phosgene is rising quite rapidly in accordance with the developments in isocyanates. Consumption in the U.S.A., which was 156 000 long tons in 1966, rose to some 268 000 long tons in 1970. The usage for 1970 was given as follows:

Toluene diisocyanate	62%
Other isocyanates	23%
Carbamate pesticides, carbonates and specialities	7%
Carbaryl	5%
Polycarbonates	3%

The isocyanates may be made from toluene (toluene diisocyanate) or an aniline-formaldehyde condensation product (diphenylmethane diisocyanate). Some 90 per cent of isocyanates finds application in polyurethane foams.

The reaction of olefins with carbon monoxide and water has also been the subject of a series of developments. The tertiary saturated monocarboxylic acids formed are commonly called Koch acids, after the originator of the process. Commercially they are offered as Versatic or Neo-acids. The olefin is reacted in the presence of a strong acid catalyst at 20–100 atm. pressure and 0–30 °C with carbon monoxide and water. A modification of the Koch process uses formic acid as the reaction medium in the place of carbon monoxide and water. Such a reaction operates at low temperature (0–25 °C) and without the need for an elevated pressure. Here again a strong acid catalyst is employed.

The Koch acids find applications in plasticizers, paint driers and synthetic lubricants.

Chapter 12

Petroleum Aromatics

Factors Affecting Supply

In Chapter 3, reference was made to catalytic reforming as a source of liquid fractions containing a high proportion of aromatics.

Aromatic hydrocarbons have been raw materials of importance to the chemical industry for many years, and their potential availability from certain petroleum processes has also been obvious for an extended period. Until fairly recently, supplies of major aromatic compounds derived from coal carbonization were sufficient to meet all chemical needs. The economics of separating individual aromatic compounds from petroleum concentrates was frequently then unfavourable, bearing in mind the value of aromatics within the petroleum industry as components of automotive and aviation fuels.

The position has been changing over a number of years. In the U.S.A. a substantial proportion of the total aromatics availability has been recovered from petroleum sources for some considerable time, as indicated by the following Table, quoting aromatics production in '000 U.S. gallons.

	BENZENE		TOLUENE		XYLENE	
	Coal	*Oil*	*Coal*	*Oil*	*Coal*	*Oil*
1954	165 000	92 000	36 000	123 000	10 000	100 000
1958	144 000	142 000	32 000	207 000	9 000	191 500
1962	114 000	432 000	27 000	334 000	8 000	347 000
1966	116 000	834 000	23 000	564 000	6 000	432 000
1969	102 000	1 084 000	20 000	740 000	5 000	377 000
1970	100 000	1 150 000	19 000	667 000	5 000	449 000

Perhaps the most significant trend is that of benzene. Fifteen years ago the coal industry provided more than half the U.S. benzene supply. Today benzene is essentially petroleum-derived in the U.S.A.

The position is governed by economic factors, and more particularly by the factor of availability. Benzene, toluene and xylenes are available from coal to the extent that coal is carbonized. This, in turn, is related to the progress of the steel- and gas-making industries. These industries are not expanding in line with the demand for aromatic hydrocarbons, and the position is further aggravated by improvements in technique, which have the cumulative effect of requiring less coal to be carbonized per unit of production in these industries.

It is not usually necessary to recover the aromatics from coal carbonization, and sometimes it may be preferable to leave some aromatic compounds in the gases from coal carbonization, where they have some value as fuel, rather than undertake the cost of their recovery. In the petroleum industry there also exists the capacity for producing aromatic concentrates. Circumstances can arise whereby these facilities are marginally surplus to petroleum requirements. The trend to jet aircraft engines, for example, reduced the demand for aviation gasoline. In these circumstances, it will readily be seen that economic factors have tended to move in favour of the petroleum source.

A complication is that aromatics recovered from catalytic reformers arise in quite the wrong proportions to meet the market demand. The pattern of supply and demand for major aromatics is indicated by the following relative outline (in percentages):

	Market demand	*Production from catalytic reforming*	*Production from coal tar*
Benzene	58	11	80
Toluene	23	55	15
Xylenes	19	34	3

It is apparent that were the catalytic reforming operation to meet benzene demand, it would throw up an embarrassing surplus of toluene and the xylenes. This has led to the development of the hydrodealkylation processes (described later in this chapter) to produce benzene from its higher homologues. These processes also present an alternative route to naphthalene.

As a result of this situation the pattern of supply for benzene in the U.S.A. was recently quoted as follows:

	%
Petroleum refinery catalytic reformers	60.4
Hydrodealkylation processes	15.6
Ethylene plant drip oils (i.e. cracked gasoline)	9.2
Coal tar and coke ovens	7.4
Imports	4.4
Drawn from inventory	3.0

The changes in benzene inventory from year to year in the U.S.A. are quite significant, and easily overlooked.

The European situation is following a rather similar pattern, on a somewhat different time scale. The practice in Europe of producing ethylene by naphtha cracking gives rise to a considerable volume of cracked spirit (or gasoline). This tends to contain its aromatics in a proportion more like the demand pattern, with benzene predominant. It is quite common for aromatics extraction from this cracked gasoline to be limited in scope to the extraction of the benzene. The toluene and xylenes in Europe have come largely from petroleum for a considerable number of years. The position of benzene has been changing abruptly as is indicated in the following table of production in western Europe:

	1965	*1970*	*1975* (est.)
Benzene produced from coal	700 000 tons	700 000 tons	500 000 tons
Benzene produced from petroleum	400 000 tons	2 100 000 tons	4 500 000 tons

If the 1975 estimate proves to be accurate the proportion of benzene derived from petroleum will have risen from 36 per cent to 90 per cent in a single decade.

Amongst the aromatics, naphthalene is a rather special case. The demand for naphthalene is essentially governed by its use as raw material for phthalic anhydride, which consumes 75 per cent of the total. The limited availability of naphthalene from coal tar sources gave rise, firstly to the development of processes for petroleum-based naphthalene and secondly, to the use of o-xylene as an alternative raw material for phthalic anhydride.

In the U.S.A. naphthalene was first produced from petroleum in 1961, and petroleum naphthalene accounted for little short of half the U.S. naphthalene supply in the middle 1960's. The economics of this operation have not proved very favourable. In 1970 coal tar naphthalene production was almost exactly 60 per cent of the total U.S. production of 317 000 long tons. The supply is completed by a modest import of around 11 000 tons. This total of 328 000 tons represents a drop of around 46 000 long tons from the peak U.S. naphthalene availability in 1968.

The production of naphthalene in western Europe fluctuated between 308 000 and 343 000 long tons per annum between 1965 and 1970, with the trend slowly but definitely downward. There is a significant net export of naphthalene from western Europe notably from west Germany.

Japanese naphthalene production rose rapidly during the 1960's from under 50 000 long tons in 1960 to a peak of about 130 000 tons in 1970. Here, too, there are signs of a diminution in production in 1971.

Recovery Processes

The recovery of aromatic compounds from the concentrates produced in such operations as catalytic reforming may be achieved by a variety of alternative techniques considered separately below.

The recovery and separation of petroleum aromatics offers a splendid illustration of the various techniques (some of which were outlined in Part One of this book) available for product separation. One aspect of this problem is represented by the separation of the aromatic compounds from the non-aromatics. A second aspect is the separation of the aromatic compounds from each other. A particular instance of the latter case arises from the development of commercial requirements for each of the xylene isomers.

(a) *Fractional Distillation*

This can seldom be used alone for the separation of pure compounds, but may be applied to the preparation of narrow-distillation-range concentrates. Not only are the aromatics difficult to separate in a pure form by virtue of the presence in the same fraction of other hydrocarbons with very similar boiling points, but they also have a tendency to form azeotropes with certain paraffins and naphthenes. In the sequence of operations whereby aromatics obtained in the catalytic reforming operation are selectively extracted (e.g. by solvent extraction) to a stage where non-aromatics are virtually excluded, the normal processing technique involves simply a clay treatment, to remove unsaturates, followed by a straight fractional distillation in three columns to separate benzene, toluene and xylene fractions. It will frequently be necessary to purify these fractions by some of the techniques outlined below.

Another instance where fractional distillation has been highly developed is in the isolation of pure ethyl benzene from a C_8 aromatic stream. The aromatic stream is derived from a catalytic reformate by solvent extraction (to concentrate the aromatics) followed by normal fractionation to separate the C_8 component. The C_8 aromatic stream may contain 20–25 per cent of ethyl benzene. This is subjected to an exceptionally close fractionation (sometimes termed superfractionation) in a 600-foot column (operated as three 200-foot columns in series) containing 350 plates and using a reflux ratio of between 25 and 50 to 1. A product purity of 99.7 per cent ethyl benzene has been claimed. This type of process is responsible for a significant proportion of the total ethyl benzene availability in the U.S.A. As is clear from the comment on isomerization processes for the xylenes, it is necessary for the operation of certain of these processes for the ethyl benzene to be removed first.

In addition, it is possible to separate o-xylene from the other xylenes by fractional distillation. It is appropriate to use about 120 plates in the column, with a reflux ratio of between 5 and 8 to 1.

(b) *Solvent extraction*

Current practice has emphasized the importance of solvent extraction to separate a high concentration of aromatics, followed by a fractional distillation to separate the individual aromatic components. The highly selective nature of the solvent systems now used for aromatics separation has reduced the complication of subsequent distillation. The selectivity of the solvent chosen for aromatics in the presence of non-aromatics should be such that a two-phase system is obtained at a convenient temperature, and that the phases should separate easily. The solvent should also be non-corrosive, non-reactive and thermally stable.

Amongst the traditional solvents used for the separation of aromatics from petroleum fractions are sulphur dioxide and furfural. In modern practice, where the main purpose of the operation may be the recovery of the aromatics as pure compounds, the commonest technique is the Udex process, using aqueous diethylene glycol as the solvent. The mixture contains about 10 per cent of water and the process is operated at a temperature of around 150 °C. After the extraction, the mixture of aromatics is passed through a contact clay treater, and then to the distillation train.

In the Udex process itself some flexibility in the solvent medium has been indicated. It has been suggested that a 3 : 1 mixture of diethylene glycol and dipropylene glycol may be used with advantage. Tetraethylene glycol has also been recommended as a component for such a solvent medium.

The range of selective solvents used for the separation of aromatics extended considerably during the 1960's. Sulfolane (see Chapter 9) has been used, sometimes in equipment previously designed to use aqueous diethylene glycol. Such a change has been claimed to increase the effective capacity of an existing unit by up to 25 per cent.

In Germany, much attention has been paid to the use of N-methylpyrrolidone as the selective solvent. The process may take several forms, including the recovery of a whole range of aromatic hydrocarbons (the Arosolvan process) or the recovery of a single component, usually benzene (the Distapex process).

A similar type of development in France involves the use of dimethylsulphoxide as the solvent medium.

Another recent contender in this process field is the Morphylane process which uses a morpholine derivative (such as N-formylmorpholine, a reaction product of formic acid with morpholine). This is commonly a combination of liquid–liquid extraction and extractive distillation. This has been commercialized in Germany. It has been quoted as particularly applicable to the extraction of aromatics from hydrogenated pyrolysis gasoline (i.e. the liquid fraction obtained from naphtha cracking after it has been treated to minimize unsaturation).

Solvent extraction is by far the most common means of recovering benzene and toluene in a pure state.

(c) *Extractive Distillation*

This process involves distillation in the presence of a solvent which becomes the least volatile component of a mixture. The solvent is selected to produce an increase in the relative volatility of the components to be separated. Requirements for a satisfactory solvent include: non-corrosivity to equipment, non-reactivity with feed components, thermal stability, high selectivity and a higher boiling point than feed components (to enable ease of separation from them). A suitable extractive solvent for benzene or toluene recovery is phenol.

Such a process is applied to a fraction in which a specific aromatic component is present, together with non-aromatics in a narrow boiling range. After the extractive distillation the aromatic compound (normally benzene or toluene) is given a clay treatment or acid treatment to remove residual unsaturates and improve the colour.

(d) *Azeotropic Distillation*

This involves distillation in the presence of a solvent which becomes the most volatile component of a mixture. The solvent is selected, as in extractive distillation, to produce an increase in the relative volatility of the components to be separated. The choice between azeotropic and extractive distillation is an economic one, depending upon the concentration of aromatics in the feed. Azeotropic distillation is generally favoured for a feed containing upwards of 40 per cent aromatics. The main difference in practice is that in azeotropic distillation the added agent goes overhead with the non-aromatics, whereas in extractive distillation it mostly goes with the aromatics as column bottoms.

As with extractive distillation, azeotropic distillation is applied to fractions in which a specific aromatic component is present, together with non-aromatics in a narrow boiling range.

Azeotroping agents which have been found effective include methyl ethyl ketone and water in combination, methyl alcohol, and nitromethane.

(e) *Solid Adsorption*

The selective adsorption of aromatics on silica gel has been employed in the Arosorb process. The adsorption in the silica gel bed proceeds to about 70 per cent saturation. Non-aromatics are washed from the bed with a volatile paraffin, such as butane, and the aromatics are then extracted by a stream of xylene. Benzene and toluene are then purified by distillation. This operation is of little significance today.

Renewed interest in this type of separation has been aroused by the Parex process. This is a continuous process using solid adsorption to separate p-xylene from a mixture of C_8 aromatics, or even the C_8 reformate fraction before extraction, and therefore containing non-aromatics.

Although this process uses a single fixed bed of adsorbent, the inlet points for the feed and for the desorbing solvent are continuously moved, and this effectively simulates a moving bed of adsorbent. Adsorption takes place at 120–175 °C and moderate pressure in the liquid phase. The pore size of the adsorbent is sufficiently large to admit any of the hydrocarbons likely to be present. The separation takes place by virtue of the greater affinity of the adsorbent for p-xylene than the other hydrocarbons present. The purity of the p-xylene product is said to be 99.5 per cent. The first commercial units operating this process came on stream during 1971.

(f) *Crystallization*

Crystallization has become significant in two types of operation—the crystallization of benzene from an aromatic concentrate, and in processes for separation of xylene isomers.

A potential source of benzene is the cracked spirit produced in plants cracking naphtha to make ethylene. In this country a process is commercially operated in which a benzene concentrate, prepared by a preliminary distillation from the cracked spirit, is chilled with refrigerated brine until benzene crystals separate out. These are removed by centrifuging.

The main application of crystallization techniques to aromatics recovery is in the separation of the xylene isomers. The boiling points of the xylene isomers and of ethyl benzene are all close together, and it is particularly difficult to separate the meta- and para-isomers. On the other hand, the freezing points of these compounds show a much wider spread.

There exists a range of such processes involving a crystallization step. They are essentially aimed at the separation of the pure p-xylene isomer. The feedstock will always contain m- and p-xylene, with the frequent addition of o-xylene and ethyl benzene. The extraction processes are commonly used in conjunction with an appropriate isomerization process.

The processes all start with a feedstock drying step. The feed is normally passed through beds of alumina or silica gel. Distillation has also been used as a technique for drying. The water content needs to be reduced to about 10 parts per million of feed. If water is present, it will cause plugging of centrifuges and rotary filters, at the low temperatures employed in subsequent processing.

The major differences in the individual processes lie in the mechanics of the crystallization and separation facilities. The feed is cooled to about −40 °C using propane and ethylene. The chilled feed passes to the first stage crystallizer which operates at −62 to −66 °C. These units may take the form of scraped surface tubular heat exchangers or tank crystallizers. The crystals formed are relatively small, and the size must be carefully controlled, so that the centrifuges and rotary filters used for the separation may be appropriately designed. The efficiency of the solid–liquid separation devices employed at this stage has been notably improved in recent years, and continuous-solid bowl centrifuges may be used.

It may be noted that the recovery of the p-xylene present is limited to around 60 per cent in view of interference by eutectic formation. A lower crystallization temperature will cause other xylene isomers to crystallize out and contaminate the product.

Crystals separated out in the first stage are partially melted and recrystallized at about −31 °C. The crystals formed at this stage are modified in shape to a more cylindrical form. The mother liquor is reduced in viscosity and drains away from the crystals quite readily. The crystals separated at this stage are about 99.5 per cent pure p-xylene. The mother liquor filtered off these crystals is recycled to the first stage crystallization.

A modification, not yet at the commercial stage, is the Institut Français du Pétrole process, which operates only one stage of crystallization. The xylene stream is made to flow in a countercurrent direction to an immiscible coolant stream. The crystals of p-xylene formed are carried with the coolant to a filter/washer, from which they pass to a melter, whence a stream of pure p-xylene emerges.

The mother liquor remaining from the p-xylene separation, and which is normally rich in unwanted m-xylene, is treated by an isomerization process in order to produce additional p-xylene. Alternatively this residual liquor may be sold as a mixed xylene solvent or used as a gasoline component.

There are two types of isomerization process. One employs a noble metal such as platinum on a silica–alumina support as a catalyst, and the other prefers to use a non-noble metal on a similar support.

The platinum-based catalyst is effective in converting ethyl benzene to xylenes. It operates in the presence of hydrogen and requires relatively infrequent regeneration. Typical operating conditions include temperatures around 450 °C and pressures in the range 10–25 atmospheres. Such processes include Octafining and Isomar.

The Isoforming process is based on a non-noble metal catalyst, but still uses elevated pressure and a hydrogen environment. The temperature of operation is in the range 370–450 °C. This process does not isomerize ethyl benzene but minimizes the need for catalyst regeneration.

The other processes using non-noble metal catalysts operate at atmospheric pressure and at temperatures of 400–500 °C. There is in general a need for multiple reaction systems as coking occurs on the catalyst, so that frequent regeneration is necessary.

The isomerization may typically increase the p-xylene content of the residual liquor from the crystallization step from its original 7–9 per cent to 20 per cent or above. This reconcentrated mixture is then recycled to the crystallization to recover the p-xylene.

All these isomerization steps operate in the vapour phase. The only process of this type which operates in the liquid phase is that of Japan Gas-Chemical Co. which uses a combination of hydrogen fluoride and boron trifluoride. This process operates at a much lower temperature than the other processes—about 100 °C. It treats pure m-xylene previously separated as described in the next section. The low temperature of operation, in combination with the pure nature of the feedstock, causes the equilibrium to favour p-xylene formation, and the conditions to minimize by-products.

(g) *Separation of m-xylene*

The maximum purity of m-xylene which can normally be obtained from the filtrate residue from p-xylene crystallization is about 85 per cent. This material has been used for the production of a mixture of isophthalic and terephthalic acids (by the Amoco oxidation process —see later in this chapter). This may be regarded as a separation of the xylene isomers by indirect methods.

At one time Chevron Chemical in the U.S.A. recovered pure m-xylene by a selective sulphonation process, said to be based on the fact that m-xylene sulphonates more rapidly than the other xylene isomers.

Union Oil of California has proposed the use of a 'clathrate former' to separate m-xylene. A complex nickel compound was recommended for this purpose. The clathrate former dissolved in hot aqueous ethanolamine is mixed with the xylenes. A two-phase mixture is formed and this is chilled. Crystallization of the chilled mixture converts it into three phases: a solid clathrate (containing p-xylene), a m-xylene—rich hydrocarbon phase, and an aqueous thinner phase. The solid is separated by filtration, and the two-phase filtrate passes to a phase separator. Here the m-xylene concentrate is drawn off and purified by distillation. This has not been commercially exploited.

The only current producer of high purity m-xylene in the U.S.A. is Arco Chemical Co. whose process has not yet been disclosed.

The main development in this field is that of Japan Gas-Chemical Co. to which reference has already been made. m-Xylene is slightly more basic than other C_8 aromatics, and will form a complex with hydrofluoric acid and boron trifluoride. The C_8 aromatics (which may include ethyl benzene) are fed to a multistage extractor in which the complex is formed. A mixture of o-xylene, p-xylene and ethyl benzene is taken overhead from the extractor, and this may be separated into its components by fractional distillation. The temperature in the extractor is 0–10 °C. The complex passes to a decomposer where at very moderate temperature (obviously less than the 100 °C at which isomerization occurs) it decomposes to form pure (99.5 per cent) m-xylene. The hydrogen fluoride and boron trifluoride which pass overhead from the decomposer are recycled to the extractor. The process is said to be marked by a minimal formation of by-products.

The main significance of m-xylene is that although it has far less use as a chemical raw material than the other xylene isomers, it is, by one of those inscrutable natural laws, by far the most abundantly produced of these isomers in the process of catalytic reforming. It is, therefore, primarily of importance as isomerization feedstock to form additional o- and p-xylene.

Production Techniques

The need for processes of hydrodealkylation, whereby surplus toluene is used to make additional supplies of benzene, has been outlined at the beginning of this chapter.

The feedstock (usually toluene but in theory any alkyl aromatic) and hydrogen pass into a reaction system comprising one or more reactors in series. There are many variations in the process; temperatures may vary from 540 to 760 °C and pressures from 20 to 67 atm. The conversion per pass is 60–90 per cent, and unconverted toluene is usually recycled. Pure benzene is separated by distillation. It is normally also given a clay treatment.

Hydrodealkylation may be simply a thermal reaction. Where catalysts are used they may be of metal, or metal oxide type on silica or alumina support. The cost of the catalyst is offset by a higher reaction rate and enhanced selectivity of the reaction.

This type of reaction became of considerable importance in augmenting the benzene supply of the U.S.A. during the 1960's, and such processes are also exerting a significant influence in Europe. Benzene is a notably volatile product in the commercial sense. As toluene has a 'floor' price equivalent to its value as a gasoline component, it is found that when benzene prices are low, the hydrodealkylation benzene production will diminish. This has offered some corrective to the supply/demand imbalances. In Europe it does not seem likely that there will, in the near future, be much further development of hydrodealkylation, since there is an important source of benzene (with relatively little toluene) in the pyrolysis gasoline or cracked spirit from naphtha cracking.

The same trend towards liquid feeds for ethylene production in the U.S.A. will also provide additional sources of benzene there. This may reduce the development of hydrodealkylation in the future in the U.S.A. also, although it is clear that there will be a continued development of catalytic reforming for refinery purposes. This is one of the many issues which must remain open to some doubt pending a resolution of the major question of policies relating to lead additives in gasoline. The elimination of these additives will stimulate catalytic reforming as a means of providing high-octane gasoline components, and increase the value of toluene as an individual component of such blends.

A further complication in the aromatics field has been the introduction of disproportionation and transalkylation processes. These may be summarized very briefly as a means of converting toluene into benzene and xylenes.

A simplified definition of the terms, is that where one molecule of toluene loses a methyl group to another toluene molecule, producing, from two molecules of toluene, one each of benzene and xylene, this is called disproportionation. Where a molecule of a C_9 aromatic loses a methyl group to a toluene molecule, so that two xylene molecules are formed, this is termed transalkylation.

The Tatoray process operates at a pressure of between 10 and 50 atmospheres, using a temperature within the range 350–530 °C. The reaction takes place in the presence of hydrogen. The heavier components of C_9 and higher which may be produced are recycled. Indeed, the feedstock may consist of C_9 aromatics (produced in abundance in catalytic reforming) as well as toluene. This process may produce xylenes and benzene from toluene in a mol ratio which can vary from 0.7 to 1 as high as 10 to 1. The corresponding U.S. process, called 'Xylenes Plus', typically produces xylenes and benzene in a ratio of about 1.5 to 1. By-products from such reactions comprise small quantities of off-gas, light aliphatic hydrocarbons, and aromatic heavy ends of C_9 and higher.

The C_8 aromatic stream produced by this means contains negligible ethyl benzene, and is consequently a first-class feedstock for xylene separation and isomerization processes. A typical balance of C_8 aromatics from different sources may be quoted as follows, as percentages:

	Catalytic reforming	*Steam cracking*	*Disproportionation*	*Coke oven*
Ethyl benzene	18–22	52	nil	16–23
p-Xylene	16–20	11	26	15–17
m-Xylene	36–41	24	50	42–44
o-Xylene	20–24	13	24	15–20

At one time there were considerable expectations of the hydrodealkylation process by which naphthalene is produced from alkylnaphthalenes present, for example, in gas oil fractions from catalytic cracking, and heavy reformate from catalytic reforming. Naphthalene of an exceptional purity is obtained by this means but the economics of the operation were found to be questionable. Although it is made by this process on a significant scale in the United States, progress has not measured up to the optimistic forecasts made a number of years ago. Another factor has been the intrusion of o-xylene into the main naphthalene market—the production of phthalic anhydride.

Usage of Aromatics

Benzene has for a long time been a major organic chemical intermediate. It has widespread applications in chemical synthesis, though the three major applications for styrene, phenol and cyclohexane are overwhelmingly important.

Usage in the U.S.A. has remained relatively static in the period 1970–71 as was the case with so many products with a spectacular growth pattern in the 1960's. The usage in 1970 was about 1 250 million U.S. gallons (or about 4.08 million long tons). This should rise to about 1 650–1 750 million U.S. gallons in 1975.

The pattern of demand in the U.S.A. for 1969 and 1975 is estimated as follows:

	1969	*1975* (est.)
Styrene	42%	44%
Phenol	20%	23%
Cyclohexane	15%	16%
Aniline	4%	4%
Maleic anhydride	3%	4%
Detergent alkylate	3%	2%
DDT	1%	1%
Dichlorobenzene	1%	1%
Miscellaneous (including ethyl benzene exports)	5%	3%
Exports	6%	2%

In western Europe as a whole the demand, which was 1.4 million tons in 1965, had risen to 3.0 million tons in 1970, and the forecast usage for 1975 is as high as 5 million tons. The pattern of demand has been changing in accordance with the following Table:

	1965	*1970*	*1975* (est.)
Styrene	36%	40%	42%
Cumene (Phenol)	21%	20%	18%
Cyclohexane	14%	23%	24%
Miscellaneous	29%	17%	16%

This pattern of growth would be regarded as most satisfactory by ordinary standards, but may be slightly disappointing in comparison with some forecasts of a few years ago. Cyclohexane production, geared as it is to nylon demand has not developed as rapidly as was once expected.

Detergent alkylate cannot be expected to achieve a very rapid growth in the future. Synthetic detergents now have a large share of the total detergent market, having ousted soap to a substantial extent. The total detergent market has achieved considerable maturity, and can only hope to grow at a rather modest rate. The linear alkyl benzenes still represent the most widely used materials in the production of biodegradable detergents. In the long run, encroachments on this market by alternative products seem inevitable, more particularly as there is some indication that the benzene ring itself is an undesirable feature in the molecular structure of a product aiming at biodegradability. It is already noticeable that alcohol ethoxylates are staking a claim in this field, and alkenyl sulphonates are also possible competitors.

Aniline is one of the building blocks of classical organic chemistry, notably in the dyestuffs field. In the past year or two it has received some stimulus from new outlets in the field of polymeric isocyanates used in polyurethanes. Benzene is nitrated using the conventional mixture of nitric and sulphuric acids at 60–95 °C. Crude nitrobenzene forms as an upper layer which can be separated from the spent acid. The crude nitrobenzene is neutralized and stripped with steam; it may be further purified by distillation, or may be used directly for aniline production. This is a catalytic hydrogenation process. In the vapour phase the reaction may take place at 270 °C and marginally above atmospheric pressure using a copper–silica catalyst. An alternative is to conduct the hydrogenation in the liquid phase under moderate pressure using a Raney nickel catalyst.

Maleic anhydride is manufactured by the vapour phase air oxidation of benzene, at 400–450 °C and atmospheric pressure, over a vanadium pentoxide catalyst. The catalyst is frequently modified by the addition of other materials (including metal oxides as promoters). Maleic anhydride is now one of the faster growing benzene derivatives. Production in the U.S.A. was about 96 000 long tons in 1970 and rose above 100 000 tons in 1971. Rather more than half of this is attributable to usage in polyester resins. Another 20–25 per cent is used to make the maleic acid trans-isomer, fumaric acid. Other applications are in agricultural pesticides and alkyd resins.

The isomerization to form fumaric acid is simply achieved. In the production of maleic anhydride it is normal to recover the reactor effluent as a solution of maleic acid in water and this has to be concentrated and dehydrated to the anhydride. It is, therefore, convenient to conduct this operation on the intermediate product, maleic acid in solution. It is possible to isomerize maleic acid, in the absence of a catalyst, merely by holding it at a temperature of about 150 °C in an inert atmosphere for a long enough period. In practice the isomerization is normally carried out at a rather lower temperature (100 °C or less) in the presence of a catalyst. Among products which have been recommended as catalysts for this isomerization are hydrogen peroxide, thiourea and ammonium salts. Some fumaric acid is normally obtained as a by-product of maleic anhydride production.

The future of maleic anhydride as a benzene consumer needs to be considered in relation to the recent renewal of activity in butylene-based production of maleic anhydride in Germany and Japan (see Chapter 9).

Toluene has traditionally occupied a much less important place than benzene in the chemical industry. Although toluene is an economical and readily available raw material for chemical synthesis, its applications are still relatively limited in number. The use for hydrodealkylation considerably exceeds all other chemical applications in the U.S.A. Figures for toluene tend to be expressed, like benzene, in U.S. gallons. The production figures have been quoted on page 209, and it will have been noted that the previous steady rise has become more irregular, and that 1970 shows a significant drop. The total U.S. consumption in 1969 was estimated to be 780 million U.S. gallons (or about 2.48 million long tons).

Only about 8 per cent of the toluene present in the streams from the catalytic reforming process is recovered as such in the U.S.A. This compares with some 53 per cent of benzene recovered, and about 13 per cent of the xylenes.

In 1969, the usage of toluene in the U.S.A. was as follows:

Hydrodealkylation to benzene	54%
Return to gasoline	19%
Explosives	10%
Solvents	9%
Toluene diisocyanate	3%
Benzyl chloride	1%
Phenol	1%
Other chemicals	3%

The position in the U.S.A. is liable to considerable fluctuation with the vicissitudes of the benzene business. As has been pointed out, it is hydrodealkylation which is phased out first when there is a surplus of benzene in the U.S.A., or when the price drops below the point of economic production by this means. There is a small but significant manufacture of toluene recovered as a by-product of styrene plants.

Some figures for toluene consumption in Japan have been presented:

	1970	*1972* (est.)
Domestic consumption of toluene ('000 tons)	660	812
Usage for paint solvents	45%	47%
mixed solvents	2%	2%
terephthalic acid	4%	3%
toluene diisocyanate	4%	5%
synthetic cresol	10%	10%
hydrodealkylation	27%	25%
miscellaneous	8%	8%

In western Europe, like Japan, hydrodealkylation is a less important factor than in the U.S.A. Of some 700 000 tons used in western Europe in 1970, excluding the quantities used for hydrodealkylation which are unrecorded, about 30 per cent was for chemical synthesis (toluene diisocyanate, phenol, caprolactam) and the remaining 70 per cent mostly found solvent applications.

The Dow process for making phenol from toluene starts with the oxidation of toluene to benzoic acid. This can be done in various ways, and Dow use the reaction of toluene and air, in the presence of a cobalt salt, at 110–120 °C and 3 atm. The purified benzoic acid, in the molten form, is oxidized to phenol and carbon dioxide by air in the presence of steam and a catalyst mix of copper and magnesium salts, at 220–245 °C and atmospheric pressure:

$$C_6H_5CH_3 \text{ (toluene)} \xrightarrow{O_2} C_6H_5COOH \text{ (benzoic acid)} \xrightarrow{O_2} C_6H_5OH \text{ (phenol)} + CO_2$$

The process for making isocyanates starts with the standard type of nitration to dinitrotoluene, using a mixture of sulphuric and nitric acids. The next step is the catalytic hydrogenation of the dinitro-compounds to the corresponding diamines—again a standard organic chemical synthesis, carried out in the liquid phase and using conventional hydrogenation catalysts. The product is largely 2,4-toluene diamine, usually with a proportion (around 20 per cent) of the 2,6-isomer.

The diamine is reacted with phosgene at a low temperature (0–5 °C). It is dissolved in an inert solvent (which may be o-dichloro-benzene) and dry hydrogen chloride gas may be blown into the solution to reduce the reactivity of the free amine. The phosgenation proceeds in stages in which phosgene adds on first to one and then to the second amino group. Hydrogen chloride is removed by blowing inert gas through the system at 110–115 °C. The 2,4-toluene diisocyanate so formed is used to produce polyurethane resins in conjunction with a polyester, or, more commonly, a polyether. The appropriate polyethers represent a major outlet for propylene oxide, and reference has been made to them in Chapter 8. Polyurethane resins form the basis of the 'plastic foam' used widely today for cushioning in furniture and vehicles. Similar resins find uses as rigid foams in lightweight construction, as durable surface coatings, and in certain rubber-like materials.

A proposed process of considerable technical interest in this field is the direct reaction of dinitrotoluene and carbon monoxide. Such a reaction will proceed in the liquid phase at about 200 atmospheres and 190 °C. A homogeneous noble metal complex catalyst is required. Clearly there will be important problems of catalyst recovery and recycle to be solved. There are recently developed noble metal resin catalysts which may prove of value in this connection.

Another contender for increased toluene usage is the Snia Viscosa route to caprolactam. Caprolactam is the monomer for nylon 6, which is expanding today at a greater rate than the more conventional nylon 6/6. There is only a limited commercial application of the Snia Viscosa route so far. The first stage of this process is the oxidation of toluene by air in the liquid phase at 150–170 °C to benzoic acid, using a catalyst system which consists of unpromoted organic cobalt salts. The next stage is the hydrogenation of benzoic acid to hexahydrobenzoic acid. This proceeds at 150 °C and 10 atm. in the presence of a catalyst comprising 5 per cent palladium on carbon. The hexahydrobenzoic acid (alternatively called cyclohexane carboxylic acid) is reacted with a solution of 75 per cent nitrosyl sulphuric acid in oleum. This reactant is formed by passing an air/ammonia mixture, containing 12 per cent ammonia, over a platinum gauze at 600 °C, cooling the gases to 0 °C, and absorbing them in oleum at 50 °C. The effective reaction is

$$N_2O_3 + H_2SO_4.SO_3 \longrightarrow \underset{\text{nitrosyl sulphuric acid}}{2SO_2(OH)ONO}$$

The so-called nitrosation of hexahydrobenzoic acid takes place at 80 °C. Caprolactam is formed with the liberation of carbon dioxide. It is necessary to use a deficit of nitrosyl sulphuric acid in order to avoid the production of explosive nitroso compounds.

CH_3 (toluene) $\xrightarrow{\text{air}}$ COOH (benzoic acid) $\xrightarrow{H_2}$ CHCOOH, H_2C, CH_2, H_2C, CH_2, CH_2 (hexahydrobenzoic acid) $\xrightarrow[\text{nitrosyl sulphuric acid}]{SO_2(OH)ONO}$ CH_2, H_2C, CH_2, H_2C, CH_2, HN — CO (caprolactam)

Although this operation can claim the advantage of a cheap raw material and fewer processing steps than benzene-based processes to caprolactam (via phenol or cyclohexane), it involves the co-production of 25 per cent more ammonium sulphate than the conventional process (via cyclohexanone oxime, see Chapter 13). The caprolactam produced requires a relatively elaborate purification process.

Terephthalic acid is normally produced from p-xylene and used in the form of its dimethyl ester to make polyester fibres. The production of pure terephthalic acid from toluene has been developed on a significant scale. It is argued that if the acid in its pure form can be made competitive in cost with the ester form, the advantages of lower plant costs and higher yields will accrue to the acid user. The counter argument is that the bulk of the existing installations use the ester anyhow, requiring modification to use the acid, that the ester is easier to handle, and that it yields a higher quality fibre. These arguments are assuming gradually less validity, and there is certainly an increasing tendency to use the acid rather than the ester.

The toluene-based processes for terephthalic acid have seemed, for some inscrutable reason, to look to Germany for their initial development and to Japan for their commercial exploitation.

The Raecke process, as its first step, once again involves the oxidation of toluene to benzoic acid by any of the conventional methods. The potassium salt is formed and subjected to a temperature of about 410–430 °C under a pressure of 10–15 atm. of carbon dioxide, and in the presence of a catalyst such as cadmium benzoate. The potassium benzoate then suffers a most interesting disproportionation, forming potassium terephthalate and benzene:

$$C_6H_5COOK + C_6H_5COOK \rightarrow C_6H_4(COOK)_2 + C_6H_6$$

An alternative process is to react toluene at 70 °C with concentrated hydrochloric acid and paraformaldehyde (a polymeric form of formaldehyde) to give chloromethyl toluene. This is 'saponified' with lime and water at 125 °C and under an elevated pressure. Methyl benzyl alcohol is separated from calcium chloride solution. When this alcohol is oxidized with dilute nitric acid at 160–180 °C and 20 atm. the effluent mixture contains a solution of o-phthalic acid in nitric acid, and insoluble terephthalic acid. The latter is centrifuged and washed with water. The process also incorporates a fluid bed esterification step using vaporized methyl alcohol and an undisclosed catalyst. This process was once operated commercially in Japan, but this is no longer the case.

The xylenes have developed extensively as chemical raw materials in the past few years. The first isomer to achieve commercial significance was o-xylene. This may be separated from the other isomers by distillation, and is oxidized to phthalic anhydride.

$$\underset{\text{o-xylene}}{C_6H_4(CH_3)_2} + 3O_2 \longrightarrow \underset{\text{phthalic anhydride}}{C_6H_4(CO)_2O} + 3H_2O$$

The oxidation takes place in the vapour phase over a vanadium pentoxide catalyst at over 550 °C. It has so far always been carried out using a fixed catalyst bed, whereas naphthalene, the competitive product, can be employed in both fixed bed and fluid bed processes. A process for fluid bed oxidation of o-xylene has been claimed but is not yet commercially proven.

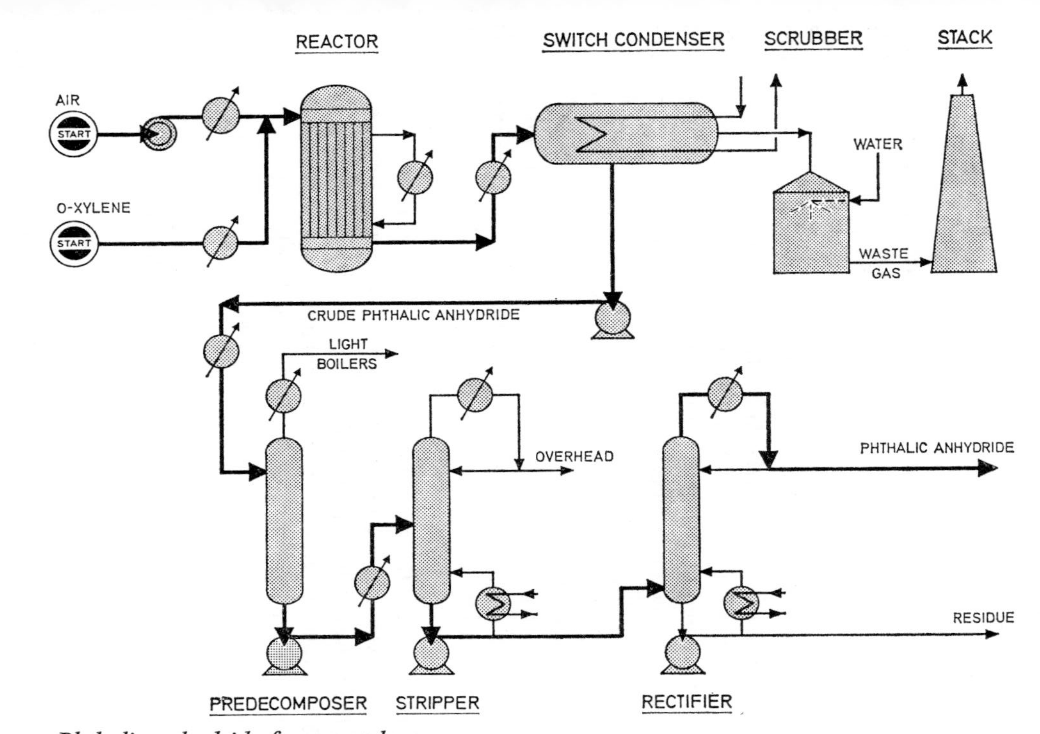

Phthalic anhydride from o-xylene

(From *Hydrocarbon Processing*, Nov. 1971)

Although this brief outline may convey the impression that not much has changed in the past few years, this process is typical of many which have been basically similar for a long time, but where a number of unspectacular improvements and innovations have made a major overall impact. Proprietary catalysts, still essentially based on vanadium pentoxide, now have a longer life. Fewer but larger reactors are used. Labour requirements have been reduced and yields improved. Many plants can operate at will on naphthalene or o-xylene feedstock.

The oxidation of o-xylene to phthalic anhydride has been carried out in the U.S.A. since 1945. Up to 1962 o-xylene supplied less than 10 per cent of U.S. phthalic anhydride. It still accounted for only about 30 per cent of U.S. phthalic anhydride production in 1970, but by the end of 1971 the capacity for this production was based 50 per cent upon o-xylene. On the other hand, about 66 per cent of west European phthalic anhydride capacity was based on o-xylene in 1970 and this capacity reached 75 per cent of the total during 1971. The Japanese picture is intermediate between these two. In 1970 about 57 per cent of Japanese phthalic anhydride was made from o-xylene. This figure is planned to rise to 78 per cent by 1975. This trend has arisen partly from the increasing inadequacy of coal tar naphthalene supplies, partly from the unfavourable economics of petroleum naphthalene, and partly from the generally favourable trend in xylene isomer separation, sparked off largely by the demands of the polyester fibre market for terephthalic acid. It may be noted that o-xylene represents a more favourable raw material stoichiometrically for phthalic anhydride than naphthalene. In practice the yield of phthalic anhydride from o-xylene on a weight basis has been only slightly more favourable than the yield from naphthalene, and there is clearly scope for much improvement.

o-Xylene production in the U.S.A. has received some buffetings in recent years. A peak of over 420 000 long tons was reached in 1968, since when there has been a drop to 369 000 tons. Of the total only about one-third has been used for phthalic anhydride production within the U.S.A. Generally around 60 per cent of total U.S. production has been exported. There are only very small markets for o-xylene other than for phthalic anhydride. A small commercial operation in the U.S.A. is the production of o-toluic acid from o-xylene.

(*Top*) Channel black plant at Skellytown, Texas

(*Bottom*) Oil furnace black plant at Sarnia, Canada

The bottom picture shows the vast improvement in cleanliness of today's oil furnace black plants compared with a typical channel black plant

Courtesy Cabot Corporation

In western Europe o-xylene production appeared in 1970 to be of the order of 115 000 long tons, which was supplemented by net imports of over 80 000 tons.

Japan, after relying to some extent on imports for a number of years, reached somewhere near a balance in 1970 with a production of nearly 70 000 long tons.

The requirement of o-xylene for phthalic anhydride manufacture in these three areas, U.S.A., Japan and western Europe, should approximately double between 1970 and 1975.

The production of phthalic anhydride in the U.S.A. in 1970 was 319 000 long tons, a rather modest increase over the figure of 301 000 tons recorded for 1966. The figure for 1970 was, however, supplemented by imports of over 20 000 long tons. The pattern of usage of this total becomes:

Plasticizers	48%
Alkyd resins	26%
Polyester resins	14%
Miscellaneous (including exports)	12%

This picture is reproduced, with slight variations, elsewhere. An interesting variant is to be found in Japan where a plant operates the Henkel process for producing terephthalic acid from phthalic anhydride. The first step in this process is to produce the dipotassium salt. Dipotassium phthalate, in the presence of carbon dioxide, and using a zinc–cadmium catalyst, isomerizes to potassium terephthalate. This is hydrolysed to terephthalic acid using sulphuric acid,

$$C_6H_4(CO)_2O \xrightarrow{2KOH} C_6H_4(COOK)_2 \longrightarrow C_6H_4(COOK)_2 \xrightarrow{H_2SO_4} C_6H_4(COOH)_2$$

dipotassium phthalate; dipotassium terephthalate; terephthalic acid

World production of phthalic anhydride was of the order of 1.5 million tons in 1970, distributed as follows:

West Europe	40%
U.S.A.	21%
Japan	14%
Eastern Bloc	12%
Others	13%

Growth of phthalic anhydride demand should be around 10 per cent a year.

The need for processes to separate m- and p-xylene arises almost entirely from the use of the latter as a raw material for terephthalic acid.

The original oxidation process for p-xylene used nitric acid under pressure at 150–250 °C. This technique is no longer favoured because of high nitric acid consumption and the formation of by-products, though variants on it are still in use. Another long established process is the Hercules–Witten process, which proceeds through several steps to the production of dimethyl terephthalate. It begins with the oxidation of p-xylene with air to give p-toluic acid, which is converted to the methyl ester. Methyl p-toluate is then oxidized with air to monomethyl terephthalate, and the latter is fully esterified with more methyl alcohol.

$$CH_3C_6H_4CH_3 \text{ (p-xylene)} \xrightarrow{\text{air}} CH_3C_6H_4COOH \text{ (p-toluic acid)} \xrightarrow{CH_3OH} CH_3C_6H_4COOCH_3 \text{ (methyl p-toluate)}$$

$$CH_3C_6H_4COOCH_3 \xrightarrow{\text{air}} HOOCC_6H_4COOCH_3 \text{ (monomethyl terephthalate)} \xrightarrow{CH_3OH} CH_3OOCC_6H_4COOCH_3 \text{ (dimethyl terephthalate)}$$

It is now more common to oxidize p-xylene in a one-step liquid phase operation. In the Amoco process the oxidation proceeds at 200 °C and 15–30 atm. using air as the oxidizing medium. The reaction takes place in the presence of a solvent such as acetic acid and a catalyst which may be a bromine-promoted cobalt salt.

$$CH_3C_6H_4CH_3 \text{ (p-xylene)} + 3O_2 \longrightarrow HOOCC_6H_4COOH \text{ (terephthalic acid)} + 2H_2O$$

This type of process has been in operation for a number of years, mainly with a view to producing a relatively crude terephthalic acid which was esterified to the dimethyl ester and purified in this form. It is, however, possible to purify the acid to fibre-grade material.

In recent years there have been many proposals for improvement in this technology. The Mobil process for pure terephthalic acid uses conditions of about 130 °C and 15 atmospheres for the oxidation of p-xylene. The reaction involves the use of pure gaseous oxygen, and it takes place in an acetic acid solvent with a soluble cobalt salt (such as cobaltous acetate) as catalyst and methyl ethyl ketone as catalyst activator. Much of the methyl ethyl ketone is converted to acetic acid. After cooling and washing, the terephthalic acid is separated by centrifuge. Most of the remaining impurities are leached out with acetic acid at elevated temperature. The terephthalic acid is separated again, washed and dried using hot nitrogen. The product, now 99.5 per cent pure, is still described as crude terephthalic acid. The acid is vaporized, treated with hydrogen and a solid catalyst, filtered in the gaseous form, and condensed as solid particles using low temperature steam and demineralized water. Residual traces of benzoic and toluic acids, and p-carboxybenzaldehyde remain in the vapours. The fibre-grade terephthalic acid has a very stringent specification indeed.

In Japan, Teijin Ltd have announced a rather similar liquid phase air oxidation. It uses a cobalt compound as catalyst and also operates in acetic acid solvent. A relatively large amount of catalyst is used, without the addition of any catalyst activator. Once again acetic acid plays a prominent part in the purification steps. At a further point there is a solution and recrystallization. It is claimed that by-product formation is minimized in this reaction so that the purification steps are simplified.

The process offered by I.F.P. is also an air oxidation of p-xylene, using an acetic acid solvent and a modified bromine–cobalt catalyst. The conditions of the reaction are said to be exceptionally mild, 180 °C and a pressure of less than 10 atmospheres.

Typically enough, a further simplification of the route to polyester fibres has been proposed. The intermediate is bis (2-hydroxyethyl) terephthalate (BHET), and the polymer is polyethylene terephthalate. It is possible to react ethylene oxide and terephthalic acid directly to form BHET.

The two reactants are pre-mixed and a catalyst is slurried into the liquid. Various amines and quaternary ammonium salts are amongst the materials mentioned as catalysts. The reaction takes place at

$$C_6H_4(COOH)_2 + 2CH_2CH_2O \rightarrow C_6H_4(COOCH_2CH_2OH)_2 \rightarrow [-O{=}CC_6H_4C{=}O-OCH_2CH_2O-]_n + CH_2OHCH_2OH$$

terephthalic acid + ethylene oxide → BHET → polyethylene terephthalate + ethylene glycol

90–130 °C and 20–30 atmospheres, and considerable agitation is applied. The product comprises a mixture of BHET and unreacted terephthalic acid, which passes through a pressure-reducing valve in the liquid form, at a temperature controlled by the evaporation of unreacted ethylene oxide. The mixed product cools to a powder form. The two components can be used together in a two-component polymerization to polyethylene terephthalate. Alternatively BHET, which is soluble in water, may be simply separated from terephthalic acid, which is not. The beauty of this operation, at least in theory, is that relatively crude terephthalic acid may be used in the reaction, and the purification may be more simply effected in the form of BHET. This process, which has reached the pilot plant stage, also involves some novel polymerization catalysts, mostly of an organometallic nature, in the place of the conventional antimony trioxide.

The most important route to polyester fibres remains the reaction of dimethyl terephthalate with ethylene glycol. In 1971 this still accounted for some 80 per cent of west European polyester fibre production. It has been suggested, however, that by 1975 fibre-grade terephthalic acid may account for 50 per cent of western Europe's polyester fibres. Some of the arguments involved in this competition were mentioned earlier in connection with the description of terephthalic acid manufacture from toluene.

In spite of a marginal reduction in U.S. production of p-xylene in 1970 (709 000 long tons), compared with 1969 (727 000 long tons), the 1970 figure still represents four times the 1965 production. Other than the production of polyester fibre and the corresponding product in film form, markets for p-xylene are extremely small. There is a substantial export from the U.S.A. accounting for something like 20 per cent of production in recent years. Western Europe, on the other hand, was a substantial net importer during the later 1960's. In 1970 the production in western Europe of around 287 000 tons compared with total usage of about 410 000 tons. The latter figure includes the p-xylene component of some significant quantities of exported intermediate products. Usage of p-xylene here had increased by 1970 to three times the 1965 figure.

Japan once again, in this area, shows the spectacular growth we have come to expect. The production in 1970 at about 260 000 tons was just about 20 times the 1965 figure.

Although it has been mentioned that, outside the fibre and film field, the markets for p-xylene and terephthalic acid are extremely small, brief mention should be made of the promising engineering plastic polybutylene terephthalate made from 1,4-butanediol and terephthalic acid. This has prospective outlets in the electronic and automotive fields. Its production was still negligible in 1970, but a U.S. production of 20 000 long tons has been forecast for 1975.

There is relatively little industrial application for m-xylene. As has already been mentioned in this chapter (under Production Techniques), according to the customarily awkward natural laws, it is not therefore surprising that it is available to the extent of almost double the figure of the other C_8 aromatics.

The separation of more or less pure m-xylene is carried out (using the techniques described earlier in this chapter) to provide a feedstock for the production of isophthalic acid. The original process for this was a liquid phase oxidation at 300–350 °C and 150 atm. in the presence of ammonium sulphate and ammonium polysulphide in aqueous solution. More recently the process took the form of a sulphur oxidation of m-xylene in an aqueous ammonia medium, at high temperatures and pressures; this yielded an intermediate amide which was hydrolysed to isophthalic acid.

The 1971 position was that there were two producers in the U.S.A. and virtually none elsewhere. The two U.S. producers are now Arco Chemical Co., whose process has not been disclosed, and Amoco Chemical Co., who use a bromine-promoted cobalt catalyst in the liquid phase—just as described for the oxidation of p-xylene. Amoco are planning a European plant in Belgium scheduled to start up in 1972.

The production of isophthalic acid in the U.S.A. has been progressing steadily and by 1970 was about 45 000–50 000 long tons.

m-xylene ($C_6H_4(CH_3)_2$) $\xrightarrow{O_2}$ isophthalic acid ($C_6H_4(COOH)_2$)

Of this about 40 per cent was used in polyester resins, 21 per cent in alkyd resins, 12 per cent in miscellaneous applications and 27 per cent went to export. This virtually represented the world supply of isophthalic acid, and needed some 35 000 tons of m-xylene.

In unsaturated polyesters, products based on isophthalic acid claim high impact strength, excellent craze resistance and good resistance also to corrosion stress. Isophthalic acid has a special application in a modified nylon tyre cord. In many outlets, products based on isophthalic acid are in direct competition with comparable products from phthalic anhydride, which is significantly cheaper. This is likely to leave isophthalic acid in the speciality class. Nevertheless, an annual growth of perhaps 8–10 per cent is to be expected.

A feature of the Amoco one-step oxidation process using bromine-promoted cobalt salts (described in relation to the oxidation of p-xylene) is that it may be applied to many aromatic compounds or mixtures of them. This, on paper, allows for the separation of the acids as an alternative to the separation of the aromatic compounds. The separation of the acids has proved quite a problem, and this technique is unlikely to see much development.

By now, naphthalene would have been in extremely short supply had its major market, phthalic anhydride, not been eroded by the switch to an o-xylene feedstock. There seems no further intention to expand petroleum-based naphthalene in the U.S.A., or to introduce it to Europe.

It has already been mentioned that some 75 per cent of naphthalene is used for phthalic anhydride, both in the U.S.A. and worldwide. In 1970 the actual figure for the U.S.A. was 72 per cent. Of the remainder rather over 10 per cent was for insecticides (the product Sevin-1, naphthyl-N-methylcarbamate), 4 per cent for dyestuffs (mostly by way of beta-naphthol), 3 per cent for synthetic tanning agents, 2 per cent for mothballs and related products, and the residual 8–9 per cent found various uses.

A brief reference should be made to the increasing industrial significance of some C_9 and C_{10} aromatics. Pseudocumene (1,2,4-trimethyl benzene), mesitylene (1,3,5-trimethyl benzene) and durene (1,3,4,5-tetramethyl benzene) are amongst the more significant. Separation processes are similar to those for the xylenes, commonly involving fractional distillation and crystallization. The importance of these materials lies mainly in the products of oxidation. Trimellitic anhydride (from pseudocumene) has applications in speciality plasticizers; trimesic acid (from mesitylene) and pyromellitic anhydride (from durene) have applications in high polymers, particularly in some of the engineering plastics ushering in the 'space age'.

Pseudocumene, in particular, is an abundant and potentially economical feedstock. Its oxidation may be carried out using nitric acid in a straightforward non-catalytic process. The reaction takes place between pseudocumene and a dilute (about 7 per cent by weight) solution of nitric acid in water at 170–190 °C and about 20 atmospheres. The reaction is exothermic and the temperature is controlled by the flow of reactants. The liquid from the reaction is cooled and partially evaporated in a crystallizer. Crystals of trimellitic acid are separated in a centrifuge and dried.

Nitrogen oxides evolved during the reaction pass to an absorption tower where they are reconverted to nitric acid.

The trimellitic acid may either be purified for use as such or heated to 220–230 °C at which point water splits off and trimellitic anhydride is formed. A semi-commercial unit of capacity 1 200 tons per year operates this process in Germany. Plans are in hand for developing it to a more substantial scale.

Chapter 13

Cyclic Compounds

In dealing with the wide range of products coming under this heading, it is necessary to be selective in picking out the relatively few materials which, as chemical entities, have achieved some measure of industrial significance in the petroleum chemical field.

The three headings below provide a brief consideration of some industrially important products and product groups with a cyclic structure, whose fields of interest are strikingly varied.

Cyclohexane

The major significance of cyclohexane, as a separate chemical entity, is its use as a raw material in the production of nylon. On the one hand cyclohexane may be oxidized directly to adipic acid. Hexamethylene diamine may be made from adipic acid by way of adiponitrile (see Chapter 9). Adipic acid and hexamethylene diamine are the reactants for nylon 6/6. Additionally cyclohexane is the major raw material for the production of caprolactam, the monomer for nylon 6.

Traditionally cyclohexane was derived from natural gasoline by fractional distillation. By this means a product of 85 per cent purity could easily be obtained (the remainder being a mixture containing methylcyclopentane and paraffins). By more elaborate forms of fractionation (sometimes termed superfractionation) a cyclohexane of 98 per cent purity could be obtained. A declining proportion, at present of the order of 15 per cent, of U.S. cyclohexane is made by this means, and the remainder by benzene hydrogenation. Elsewhere benzene hydrogenation is almost universally practised, and this gives rise to a cyclohexane of very high purity (about 99.9 per cent).

There are many modifications in the hydrogenation of benzene to cyclohexane, but the most normal basis is to operate in the liquid phase, at pressures of 20–40 atm. and using temperatures in the range 170–230 °C, in the presence of a catalyst comprising nickel on an inert support. Such a catalyst is sensitive to sulphur poisoning so that pure benzene free from sulphur is required as the raw material. At temperatures above 230 °C the thermodynamic equilibrium becomes unfavourable. Temperature control is commonly effected by some recycle of cyclohexane which acts as a cooling medium by virtue of its own vaporization. Some of these processes take the reaction only to 95 per cent completion in the main reactor and include a second, adiabatic, reactor to complete the hydrogenation.

Less commonly the reaction may proceed in the vapour phase. DSM use a noble metal catalyst at 35 atm. and a high initial temperature to carry out this hydrogenation. The temperature is decreased as the reaction proceeds in a cooled tubular reactor, so that the equilibrium mixture at the reactor tube exit will have the required low benzene content.

$$C_6H_6 + 3H_2 \longrightarrow C_6H_{12}$$

benzene → cyclohexane (CH_2 ring: H_2C, CH_2, CH_2, CH_2, H_2C, CH_2)

Another vapour phase process has been announced by Toyo Rayon Co., and this has been put into commercial operation at Kawasaki in Japan, but no technical details have been given. A vapour phase process should have the advantage of minimizing any problems of catalyst separation. The saving of energy, which should be implied by operating at lower pressures, may prove largely illusory if the hydrogen is available as off-gas from catalytic reformers already at an elevated pressure.

The main importance of cyclohexane relies upon the industrial usage of its oxidation products, starting with a mixture of cyclohexanone and cyclohexanol.

Cyclohexane has been conventionally oxidized with air in the presence of a cobalt salt (the naphthenate and the octoate have been used) at temperatures of about 140–165 °C and pressures of 8–12 atm. The reactor product is washed with caustic soda and water to remove acids and esters; the unconverted cyclohexane is separated by distillation and recycled. Some subsequent operations are based on using the 'mixed oil', containing both cyclohexanone and cyclohexanol, as feedstock. Where cyclohexanol is required (as a solvent or in plasticizer production) it may be separated and purified by distillation. More frequently it is the cyclohexanone which is required and in such a case the cyclohexanol is dehydrogenated. This has been carried out in the past in the vapour phase over a zinc–iron catalyst at 400 °C. DSM have a modified gas phase dehydrogenation process. This operation is probably more often carried out in the liquid phase. This can take place at 200 °C in the presence of a heavy solvent using cupric chromite as catalyst.

$$\underset{\text{cyclohexane}}{(CH_2)_6} \xrightarrow{O_2} \underset{\text{cyclohexanone}}{CO(CH_2)_5} + \underset{\text{cyclohexanol}}{CHOH(CH_2)_5}$$

Yields obtained in the production of 'mixed oil' from cyclohexane have risen over the years from about 65 per cent to over 70 per cent. The reaction is, however, only taken to 10–15 per cent conversion per pass.

The relatively low conversion in this operation arises from the formation of more highly oxidized by-products. An important new development in this field stems originally from the work of the Soviet chemist Bashkirov, and this uses oxidation in the presence of boric acid. Using boric acid to the extent of up to 5 per cent of the hydrocarbon, cyclohexane may be oxidized in the liquid phase at 155–175 °C and 8–10 atmospheres, with air as the oxidizing medium. The boric acid esterifies with cyclohexanol and the esters are screened from further oxidation. The esters hydrolyse easily, and the boric acid is recycled. It is necessary to remove water formed during the reaction. The partial pressure of water vapour in the reactor vapour phase must be maintained at a low level in order to obtain the desired oxidation selectivity. It is possible to take advantage of the formation of a water/cyclohexane azeotrope in order to assist this.

The particular characteristic of this oxidation is that the ratio of cyclohexanol to cyclohexanone in the 'mixed oil' is 9 or 10 to 1. The conversion per pass is limited to around 10 per cent, but the ultimate molar yield of the desired products on cyclohexane is 90–95 per cent. The improvement in yield is to some extent offset by the cost of boric acid, and a rather higher capital cost than for the conventional process.

The 'mixed oil', comprising cyclohexanol and cyclohexanone, may be oxidized to adipic acid, using 60 per cent nitric acid at 60–80 °C and elevated pressure, in the presence of a catalyst complex in which both copper and vanadium are present (e.g. ammonium vanadate and copper turnings).

It is possible to carry out the two oxidation steps from cyclohexane to adipic acid both with nitric acid, but this has now been abandoned industrially, because of the excessive usage of nitric acid, and considerable by-product formation.

A one-step oxidation from cyclohexane to adipic acid with air is possible (e.g. with cobalt acetate at about 100 °C and 11 atm.) but does not seem to be generally favoured. One of the problems of such an operation is the tendency for the formation of by-products of little value.

A two-step air oxidation of cyclohexane to produce adipic acid is coming increasingly into favour. The first air oxidation step can be either the conventional cobalt catalysed operation or may use the newer boric acid catalyst. The second step may be carried out in acetic acid solution. The catalyst may be a mixture of copper and manganese acetates. In such a case the conditions used are 80–85 °C and 6–7 atm.

Adipic acid is allowed to crystallize out and can then be separated by centrifuge. The dilute nitric acid in the mother liquor is recycled. Nitrogen oxides, which are taken overhead from the reactor, are sent to an absorption tower for recovery as nitric acid. Even so the usage of nitric acid represents a significant cost in this process.

CO		CHOH		H_2
H_2C CH_2	or	H_2C CH_2	$\xrightarrow{O_2}$	H_2C—C—COOH
H_2C CH_2		H_2C CH_2		H_2C—C—COOH
CH_2		CH_2		H_2
cyclohexanone		cyclohexanol		adipic acid

Adipic acid is one of the component monomers for nylon 6/6. As has already been pointed out, the second monomer is hexamethylene diamine. This is made by the hydrogenation of adiponitrile (Chapter

9). Adiponitrile may be made by the classical route from adipic acid and ammonia. There are competitive routes from butadiene (Chapter 9) and acrylonitrile (Chapter 8).

The production of cyclohexanone from phenol is also a well-established process to which reference has been made in Chapter 8. There is also a one-step process operated by Allied Chemical in the U.S.A. for many years, which carries out the hydrogenation using palladium on carbon as catalyst. In general, cyclohexanone is now more often made from cyclohexane. This preference has in certain circumstances proceeded to the extent of using cyclohexane oxidation as a route to phenol. The 'mixed oil', cyclohexanol and cyclohexanone, is dehydrogenated over a nickel or platinum catalyst at 400 °C. Such a process has been taken to the commercial stage but has not developed to any significant extent. This may be due to a combination of processing and economic factors. A technical point affecting this processing technique is that phenol and cyclohexanone form an azeotropic mixture (75 per cent phenol/25 per cent cyclohexanone) which cannot be separated by normal distillation. If the expense of extractive distillation is to be avoided, it is therefore necessary to establish dehydrogenation conditions so that the phenol content of the exit gas should exceed 80 per cent. This calls for a high temperature reaction.

The major alternative form of nylon to nylon 6/6 is nylon 6, which is particularly well established in Europe, and has now gained a very substantial market in the U.S.A. Nylon 6 is based on the single monomer caprolactam. There is one caprolactam process, used in Italy, based on toluene (see Chapter 12). Processes proceeding from cyclohexanone can use phenol as an alternative raw material to cyclohexanone, but the current preference is generally for cyclohexane (some 75 per cent of world caprolactam is made from cyclohexane).

Whilst the cyclohexane oxidation is likely to proceed as usual by way of producing a mixture of cyclohexanol and cyclohexanone, in this case only the latter is required. The 'mixed oil' will therefore require dehydrogenation, as discussed earlier.

What might now be termed the 'conventional' process to caprolactam proceeds by the reaction of cyclohexanone at moderate temperature (up to about 100 °C) with hydroxylamine sulphate. The mixture is neutralized with ammonia. The products from the reactor pass to a separator where cyclohexanone oxime, maintained above its 90 °C melting point, settles out as an oily liquid. The crude product containing 4–5 per cent water is subjected to the Beckmann rearrangement by treatment with 20 per cent oleum at 120 °C. The solution containing caprolactam is continuously withdrawn from the system, and promptly cooled to below 75 °C to avoid hydrolysis. It is then cooled further, and neutralized with ammonia. The caprolactam is finally recovered by distillation or by solvent extraction with benzene.

$$\underset{\text{cyclohexanone}}{\begin{array}{ccc} & CH_2 & \\ H_2C & & CH_2 \\ H_2C & & CO \\ & CH_2 & \end{array}} \xrightarrow[\text{hydroxylamine}]{NH_2OH} \underset{\text{cyclohexanone oxime}}{\begin{array}{ccc} & CH_2 & \\ H_2C & & CH_2 \\ H_2C & & C{=}NOH \\ & CH_2 & \end{array}} \xrightarrow{H_2SO_4} \underset{\text{caprolactam}}{\begin{array}{ccc} & CH_2 & \\ H_2C & & CH_2 \\ | & & | \\ H_2C & & CH_2 \\ | & & | \\ HN & - & CO \end{array}}$$

By this process, about 4.5 to 5 tons of ammonium sulphate are produced for each ton of caprolactam. It is not surprising, therefore, that in the United Kingdom development of this process, a chemical enterprise (DSM) entered into partnership with Fisons, a company with vast fertilizer interests, who are therefore in a position to dispose of the considerable tonnage of ammonium sulphate by-product.

In general terms, the co-production of ammonium sulphate with the caprolactam is a nuisance and various attempts have been made to minimize or eliminate this.

The process of the Japanese company, Toyo Rayon, proceeds directly from cyclohexane, employing an operation called photonitrosation. A series of initial steps is involved in the preparation of the reaction medium which is nitrosyl chloride in conjunction with hydrogen chloride. Cyclohexane is treated with this mixture at below 20 °C under the influence of actinic light from a mercury vapour lamp. Cyclohexanone oxime hydrochloride is formed directly in this one step.

$$\underset{\text{cyclohexane}}{\begin{array}{ccc} & CH_2 & \\ H_2C & & CH_2 \\ H_2C & & CH_3 \\ & CH_2 & \end{array}} + \underset{\substack{\text{nitrosyl}\\\text{chloride}}}{NOCl} + HCl \rightarrow \underset{\substack{\text{cyclohexanone oxime}\\\text{hydrochloride}}}{\begin{array}{ccc} & NOH.2HCl & \\ & \| & \\ & C & \\ H_2C & & CH_2 \\ | & & | \\ H_2C & & CH_2 \\ & CH_2 & \end{array}}$$

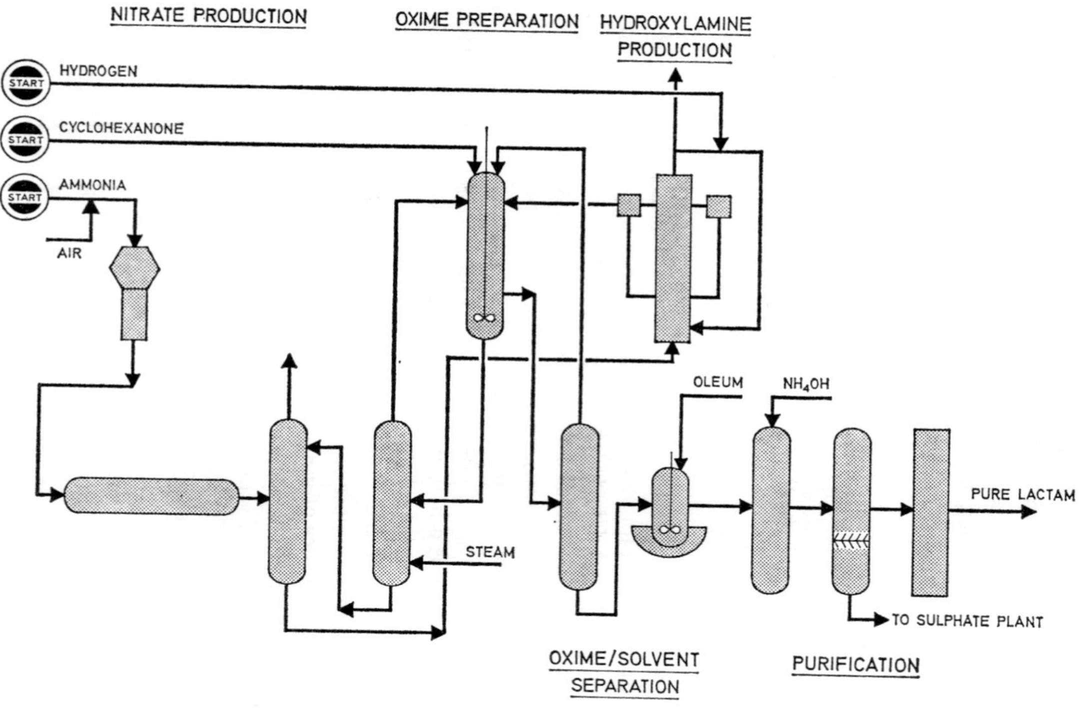

Caprolactam (low sulphate process)
(From *Hydrocarbon Processing*, Nov. 1971)

Operating at about 10 °C, and diluting the nitrosyl chloride with inert gas, minimizes by-product formation. The presence of hydrogen chloride inhibits the side reaction which would form 1,1-chloronitrosocyclohexane.

The mercury vapour lamp is kept clean by a flow of sulphuric acid over its surface. The acid is subsequently used in the Beckmann rearrangement.

The cyclohexanone oxime hydrochloride, which forms as a heavy layer at the bottom of the reaction vessel, is withdrawn and treated with sulphuric acid in the presence of some oleum. The reaction is vigorous, and temperature control is important. Hydrogen chloride is generated and recycled. The oxime then suffers the normal Beckmann rearrangement to caprolactam. On neutralization of the reaction mixture with ammonia, the crude caprolactam separates as an upper layer from ammonium sulphate solution. It is purified by vacuum distillation.

The process simplification obtained in this procedure is likely to be offset partially by the substantial power cost, and the cost of maintaining the lamp system.

A process once operated by du Pont in the U.S.A., but not apparently elsewhere, starts with the nitration of cyclohexane with 30 per cent nitric acid at 120 °C and 4 atm. to nitrocyclohexane. This is reduced to cyclohexanone oxime over a zinc–chromium catalyst (100–150 °C, 3 atm.) and the conventional Beckmann rearrangement follows. This process suffers from the formation of a number of by-products in the nitration step (notably adipic acid, which du Pont could use for nylon), and the catalyst life for the reduction step is poor. It is understood that du Pont abandoned this process in 1966–67.

Both the photonitrosation and the nitration processes succeed in reducing the by-product formation of ammonium sulphate to about half that of the conventional process via cyclohexanone oxime, since the production of hydroxylamine itself is responsible for half of the ammonium sulphate formed in that operation.

Hydroxylamine is in this instance used in the form of its sulphate. The first step in this production is a reaction between ammonium carbonate solution and nitrogen oxides directly from the catalytic air oxidation of ammonia over platinum at 850 °C. This reaction produces ammonium nitrite. An equimolar mixture of ammonium nitrite and ammonia is contacted in an absorption tower with sulphur dioxide at atmospheric pressure and low temperature. This forms hydroxylamine disulphonic acid. A double hydrolysis operation forms hydroxylamine and ammonium sulphate, and the hydroxylamine sulphate is then produced.

Another process development working on the elimination of this part of the ammonium sulphate by-product has been proposed by DSM. This uses a gas–liquid contactor in which hydrogen gas is contacted with a circulating inorganic liquid containing nitrate ions, a buffering acid, and a noble metal catalyst (based on platinum, rhodium or palladium). This forms hydroxylamine directly from the hydrogenation of the nitrate ions. A small amount of ammonium ion is formed which does not interfere with the reaction and which is destroyed later in the process. This development has been introduced into commercial units in Japan, the U.S.A. and the United Kingdom.

Toagosei have claimed an ammoxidation process for converting cyclohexanone directly into the oxime. The liquid phase reaction is between cyclohexanone, ammonia and hydrogen peroxide, in a water solvent system. The temperature used is 10–30 °C and the pressure is marginally above atmospheric. A heteropolytungstate catalyst is preferred, and a tin compound may be used as an assistant catalyst. The presence of a soluble sulphate increases the cyclohexanone conversion rate and the oxime yield.

The Union Carbide caprolactam process avoids the formation of by-product ammonium sulphate, but in this case acetic acid is co-produced with caprolactam. This process, like several others, uses cyclohexanone as raw material. The first step is the oxidation of cyclohexanone to caprolactone with peracetic acid. The peracetic acid is formed by the controlled oxidation of acetaldehyde.

The oxidation of acetaldehyde to acetic acid—a very long-established process—once made a point of using a manganese acetate catalyst specifically for the purpose of stimulating the breakdown of peracetic acid (deemed an explosion hazard) to acetic acid. By carrying out the oxidation of acetaldehyde at low temperature the initial product is acetaldehyde monoperacetate. This intermediate product readily converts to peracetic acid and acetaldehyde at 100 °C under reduced pressure. It is the low temperature liquid phase oxidation of acetaldehyde that is employed by Union Carbide.

An alternative direct oxidation of acetaldehyde in the vapour

ICI plant for making cyclohexane from benzene

Courtesy Institut Français du Pétrole

The nerve centre—a control room at the polyolefins works at Wilton

Courtesy IC

A model of the Petkim ethylene plant in Turkey—a technique used by contractors to study plant layout and piping design

Courtesy Stone and Webster Engineerin

phase with oxygen to produce peracetic acid, is employed by British Celanese in the United Kingdom. The main problems of this process are said to be the effective dissipation of the heat of reaction, elimination of explosion hazards, and the rapid separation of peracetic acid from unreacted acetaldehyde.

$$\underset{\text{acetaldehyde}}{CH_3CHO} + O_2 \rightarrow \underset{\text{peracetic acid}}{CH_3COOOH} \xrightarrow{+\text{cyclohexanone}} \underset{\text{caprolactone}}{\begin{array}{ccc} & CH_2 & \\ H_2C & & CH_2 \\ | & & | \\ H_2C & & CH_2 \\ | & & | \\ O & - & CO \end{array}} + \underset{\text{acetic acid}}{CH_3COOH}$$

The reaction of peracetic acid and cyclohexanone takes place in the liquid phase, using an inert solvent such as acetone, a temperature of 25–50 °C and atmospheric pressure.

The reaction of caprolactone with ammonia in the presence of water at 300–450 °C and 300–400 atm. sets up a complex series of equilibria, by virtue of which some 50 per cent of the caprolactone is converted to monomeric caprolactam. The latter is extracted from the reaction mixture and the by-products recycled. The ultimate yield of caprolactam is said to be 85–95 per cent on the caprolactone.

Although, as has been stated, this process avoids the formation of ammonium sulphate as a by-product, there is still a co-product which is acetic acid. A commercial plant used this process in the U.S.A. but it is understood that Union Carbide have discontinued the production of caprolactam.

Another process, which has reached the pilot plant stage, could prove of interest. Briefly, nitrocyclohexanone is formed from nitric acid and cyclohexenyl acetate (itself produced from cyclohexanone and ketene). The ring is broken in a hydrolytic cleavage to form nitrocaproic acid, and this is hydrogenated to aminocaproic acid. Aminocaproic acid may be thermally cyclized to caprolactam. The only by-product is acetic acid which may be recycled to ketene. This may, however, prove too elaborate a means of by-product elimination in the rather stern economic pattern of caprolactam.

It is clear from the process descriptions that the demand for cyclohexane is dominated by nylon applications, though marginally less than in the past. Probably about 90 per cent of world cyclohexane is destined for nylon. In the U.S.A. the suggested pattern is that 40 per cent is for nylon 6/6 (i.e. adipic acid), 15 per cent is for nylon 6 (i.e. caprolactam), 12 per cent is for miscellaneous uses, and 33 per cent goes to export. The relative importance of nylon 6 and nylon 6/6 varies considerably with geographical area. In the United States nylon 6 fibre production rose from about 20 per cent of total polyamide fibres in the mid-1960's to 28 per cent in 1970. Virtually all the remainder is nylon 6/6. In west Europe 43 per cent of total polyamide fibre production was represented by nylon 6 in 1970, and the figure for Japan in that same year was 99 per cent.

If nylon 6/6 is made entirely from cyclohexane raw material, 1.46 lb of cyclohexane is required for each lb of nylon 6/6. If only the adipic acid co-monomer is produced from cyclohexane the requirement per lb of nylon 6/6 drops to 0.65 lb.

The production of nylon 6 from cyclohexane requires about 1.1 lb cyclohexane per lb nylon 6.

About 88 per cent of adipic acid in the U.S.A. finds its application in the form of nylon 6/6 as fibre or resin. A further 6 per cent is used as an ester in plasticizers and synthetic lubricants, and 5 per cent enters into polyurethanes. The remaining 1 per cent finds miscellaneous uses. Earlier hopes of applications as an acidulant for foodstuffs have not so far been justified.

Amongst other cyclohexane derivatives, cyclohexanone has some solvent and chemical applications, and cyclohexanol is used in the form of cyclohexyl esters. Cyclohexane itself is used as a solvent in the production of high density polyethylene according to the solution form procedure of the Phillips process. The export of cyclohexane from the U.S.A. is a vulnerable market, and some indication of the recent reduction in this is given later.

Caprolactam has some minor applications outside the nylon field, in a variety of formulations. Of more chemical interest is the process by which lysine may be made from caprolactam in a four-step synthesis. Lysine is an amino-acid used as a cattle feed supplement, and it is normally made by a fermentation process. The synthesis route was pioneered by DSM in the Netherlands, but the operation of the commercial unit appears to have been discontinued.

The process for making synthetic phenol from cyclohexane does not, at this stage, seem likely to develop into a major factor. The 'mixed oil', comprising cyclohexanol and cyclohexanone, has also been suggested as a raw material for aniline, whereas cyclohexylamine may be derived from cyclohexanol.

Cyclohexane has not had a long history of industrial significance. It is only since 1950 that it has become commercially important. More recently the growth freely predicted in the middle 1960's has failed to materialize.

U.S. production is frequently expressed in terms of millions of U.S. gallons (and each gallon of cyclohexane amounts to about 6.5 lb). Production of cyclohexane in the U.S.A. in 1966 exceeded 750 000 long tons. The 1970 figure was around 820 000 tons but this fell to about 680 000 tons in 1971. A major factor in this was the fall in the export market. This amounted to about 360 000 long tons in 1968, but had dropped to under 250 000 tons by 1970. In spite of this, the underlying trend remains upwards, and the total U.S. demand (which includes the export demand) could reach 1 million tons in 1975.

In Europe, too, the expansion in cyclohexane, while substantial, has not met all the optimistic forecasts. Production in west Europe in 1970 was marginally over 600 000 long tons and the figure forecast for 1975 is about 1.1 million long tons.

Cyclopentadiene

Cyclopentadiene is found in the C_5 fraction obtained in the vapour phase cracking of naphtha, and also in the light oils derived from coal carbonization. Cyclopentadiene can readily be separated by the ease with which it can be dimerized by 'soaking' at a slightly elevated temperature. The dimer can readily be depolymerized back to cyclopentadiene.

In the U.S.A. the usage of cyclopentadiene is rather static at about 25 000 long tons per year. Of this total some 80 per cent is for pesticides.

Cyclopentadiene undergoes a complex Diels–Alder reaction with acetylene and hexachlorocyclopentadiene to give the product aldrin. Dieldrin is the epoxy derivative of aldrin. Other products in the same family are endrin and isodrin. These are very effective but very persistent pesticides. Whilst they have many virtues, including the capacity of dieldrin to attach itself to wool like a dyestuff and protect the fibres from moths and other pests, their very persistence is claimed in some circumstances to be an environmental hazard. The use of these products is consequently subject to a variety of restrictions.

The only other application of any note at the present time for cyclopentadiene is its use as a termonomer in ethylene–propylene termonomer rubbers. This application has so far proved rather disappointing.

The proposals to recover isoprene from the C_5 fraction of pyrolysis gasoline (produced in naphtha cracking plants) require that the cyclopentadiene content be first dimerized and separated. This would throw up a substantial source of cheap cyclopentadiene and may awaken interest in new derivatives. One possibility lies in the elastomeric polymers of cyclopentene, which is made by hydrogenation of cyclopentadiene.

Naphthenic Acids

Naphthenic acids occur naturally in crudes from certain areas (notably Venezuela, Trinidad and California). They are carboxylic acids of substituted cyclopentanes and cyclohexanes, with a wide range of molecular weight from 114 upwards.

The acids of major commercial importance are separated from gas oil fractions in the boiling range 200–370 °C. They are extracted with dilute sodium hydroxide solution, and regenerated by acidification with sulphuric acid. Appropriate purification and distillation treatments are given according to grade. The molecular weight of the commercial products covers the range 180–350.

The usage of naphthenic acids is lower in the U.S.A. today than in the middle 1960's. In 1965 about 18 000 long tons were used. This had dropped to under 15 000 tons in 1969, and a forecast for 1973 is only marginally above 15 000 tons.

Of this usage, about 75 per cent is for paint driers, fungicides, catalysts and additives for high pressure lubricating oils. A further 9 per cent is for corrosion inhibitors, and the remaining 16 per cent includes miscellaneous uses and exports.

In a number of applications alternative products are attacking the naphthenic acid market. These include Koch acids (made from olefins, carbon monoxide and water—see Chapter 11), and tall oil acids.

These acids are used in the form of metal salts. Lead, manganese, cobalt and zinc naphthenates are used as paint driers. Copper and zinc naphthenates are fungicides. Alkali metal naphthenates are specialized emulsifying agents. Lead, barium and magnesium naphthenates are used as fuel oil and lubricating oil additives. Perhaps the best prospect for development of the naphthenates in the future is in the form of oil-soluble catalysts for chemical synthesis.

Chapter 14

Carbon Black

The subject of carbon black is almost a study in itself. It consists essentially of fine carbon particles, semi-graphitic in structure. The importance of this commodity may be indicated briefly by stating that 30–35 per cent of the rubber in a tyre consists of carbon black. Of all the carbon black used, a very high proportion finds application in rubber technology, as the following outline of usage in the U.S.A. indicates:

Elastomers	95.0%
Printing inks	2.2%
Plastics	1.0%
Paint	0.5%
Carbon paper	0.2%
Miscellaneous	1.1%

There is also an export from the U.S.A. which has been of the order of 10 per cent of the domestic usage. This proportion is tending to decrease.

In the channel process for carbon black, natural gas is burned so that the flames impinge upon cooled metal surfaces (commonly of 8–10 inch channel iron). The air supply is controlled so that a luminous smoky flame is produced. The temperature of combustion, 1 000–1 200 °C, is reduced to 500 °C at the metal surface, so that carbon black is deposited on the channel irons, which move backwards and forwards on overhead rails. The carbon black is scraped off into large collecting hoppers. It is then screw-conveyed into a pneumatic conveyor, which carries it via a grit separator to a cyclone collector. The black at this point has an apparent density of 4 lb/cu. ft. Where the black is to be used in the ink industry, the density is increased by agitation to about 12 lb/cu. ft, but where it is destined for the rubber industry it is pelletized (using either a wet or dry process) to an apparent density of 20–25 lb/cu. ft (for more economical storage and handling).

Variants in the channel process usually relate to the type of depositing surface. This may involve (apart from the moving channel irons) rotating discs, stationary plates or revolving cylinders.

The thermal process is based on the straight thermal decomposition of an appropriate raw material, which is normally natural gas. The furnaces operate on a cyclic basis. Refractory bricks are heated to 900–1 400 °C during the heating run, and the feed gas is then admitted. The make of carbon black continues until further heating is necessary. The gases from the cracking operation convey the thermal black to a separating system, comprising cyclone separators and bag filters. The filtered gas may then be recycled to operate the heating run of the furnace cycle. Since the recycle gas consists mostly of hydrogen, it may be used for other chemical purposes, including ammonia production. Only a proportion of the thermal black flows with the gas stream to be separated as described. The remainder, which may be half the total make, stays deposited on the furnace brickwork and ignites during the next heating run. The thermal black is a much coarser product than channel black.

The furnace process was originally developed on the basis of a natural gas feedstock. This comprises a combined partial oxidation and cracking process, taking place at temperatures around 1 300 °C. The carbon is formed in the furnace, and carried by the gas flow to a quench tower, where the temperature is reduced to about 200 °C by a water quench. The black is collected in a system comprising electrostatic precipitators, cyclone separators and bag filters. The significance of this process derives from the yield of carbon black, which, per unit of natural gas, is nearly twice that of the channel process. The product is slightly coarser than channel black.

The oil furnace process, which dates from 1943, is the most important process today. The raw material is a highly aromatic liquid hydrocarbon fraction, with properties covered by a complex and detailed specification. Much of the liquid feedstock used in the United Kingdom is imported. The principle of the furnace operation is similar in both the oil and gas furnace processes. The oil feedstock enters the furnace as a spray, atomized by air or pressure. Auxiliary gas burns to completion in the furnace with air, raising the temperature to about 1 400 °C. The quench-collection system is similar in both oil and gas furnace processes. A feature of the oil furnace process is that its dependence on gas is slight enough to make it possible for operation to take place almost anywhere, by shipment of the appropriate oil fraction. Where natural gas is not available as an auxiliary, refinery gas is commonly used. About 1.8 lb of the liquid feedstock is required per lb of carbon black produced. Some 84 per cent of all carbon black made in the U.S.A. is based on a liquid feedstock.

Passing mention should also be made of acetylene black, obtained by the pyrolysis of acetylene. This is made on a very small scale, but in view of its superior electrical properties, has found application in dry cells.

About 89 per cent of U.S. carbon black capacity in 1971 operated the furnace process (mostly using oil feedstock), nearly 10 per cent was for thermal black, and a mere 1.5 per cent for channel black.

The most important characteristics of a carbon black are pH, specific surface (governed by particle size) and particle structure. Channel blacks exhibit an acidic surface reaction, whereas furnace blacks give an alkaline reaction, arising from evaporation deposits of the water used for cooling. This difference has an important effect on the vulcanization process of the rubber compound containing carbon black.

A high specific surface (or a minimum particle size) of a carbon black imparts maximum abrasion resistance to the rubber, but the processing characteristics are less favourable.

Comparing particle structure, thermal blacks with large individual particles are described as having 'low structure', whereas oil furnace blacks, having an agglomerated chain-like structure in their particles, are characterized as having 'high structure'.

The combination of high specific surface and high structure in certain grades of oil furnace black offers a combination of good reinforcing action with the greater ease of dispersion and processing, together with durability, arising from the high structure. A disadvantage of high structure is the heat build-up which can arise in tyres. One of the developments in the gradual replacement of channel black in most fields has been the low-structure oil furnace blacks which may, for example, be deliberately induced by adding metallic sodium to the flame.

Thermal blacks are economical materials where reinforcement is not the primary need.

Demand for carbon black in the U.S.A. was about 1 291 000 long tons in 1970 and 1 330 000 long tons in 1971. The figure forecast for 1975 is about 1.5 million tons.

Carbon black production is now to be found in all major industrialized areas. Japan used 280 000 tons of carbon black in 1970 and the forecast for 1975 production is 450 000 tons. The production in Italy in 1969 was 104 000 tons.

It is clear that the U.S.A. is still by far the largest producer of carbon black. The U.S. companies in this field have maintained at least a share of the ownership in many of the plants located elsewhere.

Capacity continues to increase in line with the expanding needs of the motorized community. In the mature situation of the U.S.A. growth of carbon black production is now a sober 3 per cent per year. Elsewhere the growth is more rapid. Most new capacity is based on the oil furnace process.

Channel black originally began to lose its share of the carbon black business on economic grounds, since it is wasteful of the natural gas used. Increasing importance has been attached, with the passing of time, to the dirt and pollution arising from these plants. It seems possible that by the end of 1972 the two remaining U.S. channel black plants may have to close in accordance with Texas anti-pollution laws. This is in spite of the fact that for some specialized applications there is still no complete replacement for channel black.

The channel blacks were superior to early furnace blacks in the reinforcing and abrasion-resistant properties which they could impart to the tyre rubber. The more recent oil furnace blacks, in combination with synthetic rubber (SBR grade), have given very good road wear characteristics to tyre treads. Other types of black are used where ease of processing is more important than the reinforcing effect, e.g. for rubber mountings and for moulded and extruded goods.

Chapter 15

Sulphur and Sulphuric Acid

Traditional sources of sulphur have been the elemental sulphur recovered by the Frasch process from the domes of the U.S. Gulf Coast, and pyrites from a variety of sources. Pyrites is economic only in reasonably close proximity to its source, and the majority of the sulphur entering world trade used to be in the form of the Frasch product. This process melts the sulphur underground with hot water, and the molten sulphur is blown to the surface with compressed air. There was a critical period in 1950–52 when it was believed that this source of sulphur was to be very short-lived. Since that time, the resources of Frasch sulphur have increased to a marked extent, including the discovery of important new deposits in Mexico. The period of doubt stimulated investigation into alternative sources of sulphur, and it was found to be economic in a number of cases to recover sulphur from refinery gas or natural gas. The potential availability of sulphur is no longer a major problem. The proven reserves of sulphur resources were estimated in the late 1960's at 1 039 million long tons (in terms of sulphur). These are supplemented by a further 3 624 million tons of probable reserves. Some 27 per cent of the proven reserves and 73 per cent of the probable reserves are from a petroleum source.

These figures can be compared with world production of sulphur in all its forms, which amounted to 41.8 million long tons in 1970. Of this total, 22.9 million tons are consumed as elemental sulphur.

Some fuller details for the world as a whole have been indicated as follows (in millions of tons S):

	1957	*1960*	*1966*	*1970*
Production of sulphur in all forms	17.6	22.4	31.3	41.8
Production of elemental sulphur	8.9	10.6	16.1	22.9
Sulphur recovered from petroleum sources	1.4	2.8	5.7	9.5

During recent years the supply and demand of sulphur in all forms have been closely related, at least in the western world. In the late 1960's the consumption of sulphur outside the communist countries was marginally higher than the production, but in 1969 a balance was reached, and in 1970 there was a surplus production of 800 000 tons. Consumption of sulphur in the communist countries is not exactly known.

The geographical source of petroleum-based sulphur (essentially all recovered from 'sour' natural or refinery gas) is as follows (in million tons S):

	1966	*1967*	*1970*
Canada	1.7	2.0	4.4
France	1.5	1.5	1.7
U.S.A.	1.2	1.3	1.3
Others	0.6	0.7	2.1

The quantities under 'others' include a modest amount recovered in most industrial countries as a result of desulphurization processes in refineries.

The importance of sulphur recovery from 'sour' gases is evident enough. It may also be noted that the sour gas supplies of Canada and south-west France would be quite unsuitable for fuel or chemical use if the sulphur compounds (largely hydrogen sulphide) were not first removed.

The separation of hydrogen sulphide is normally effected by amine absorption, or equivalent processes involving additional solvents such as sulfolane, from which a concentrated stream of hydrogen sulphide can subsequently be released. This could be a source of air pollution, and it is consequently a happy solution if it is economically feasible to recover sulphur from it.

The recovery of sulphur from hydrogen sulphide is a well-established chemical process,

$$2H_2S + 3O_2 \rightarrow 2SO_2 + 2H_2O$$
$$2H_2S + SO_2 \rightleftharpoons 3S + 2H_2O$$

Hydrogen sulphide is first burned with air (and some provision, such as a waste heat boiler, is made for recovering the heat of reaction). The sulphur dioxide formed in this reaction is mixed with an appropriate volume of the original hydrogen sulphide stream, and passed through a converter (or Claus kiln) containing an activated bauxite catalyst. The gases enter the converter at 250–400 °C and leave at 400–500 °C. The converter effluent is cooled, and liquid sulphur condensed out is pumped to storage. Unconverted gases are passed to a second converter operating at somewhat lower temperatures. Various modifications exist, centred around the same principle, and usually differing mainly in the techniques used for heat recovery.

Recent developments in this field have concentrated on sulphur recovery from gas streams with a low sulphur content. The improved Claus process is in general limited to the treatment of gas streams containing 20 per cent of sulphur or more. A development in operation at Lacq in south-west France called the Sulfreen process is applied to tail gas from existing Claus units. A modified catalyst (developed by Lurgi) allows a reduction in the operating temperature. This improves the equilibrium of the second reaction forming sulphur from hydrogen sulphide and sulphur dioxide. There are four reactors out of six on stream at one time. A fifth is being regenerated and the sixth is cooling after regeneration. During regeneration the temperature is raised so that the deposited sulphur is vaporized. A stream of nitrogen sweeps the catalyst clean, while the liquid sulphur product drains to sulphur storage.

Other techniques are applied to acid gas streams containing a high proportion of carbon dioxide and a small proportion of hydrogen sulphide. Such streams arise from the operation of processes such as the Sulfinol process. One technique involves burning the gas stream with pure oxygen. There follows a two-stage catalytic converter, sulphur demisters, and finally a catalytic after-burner. In dealing with gas streams containing 5 to 10 volume per cent hydrogen sulphide, the sulphur may be recovered to the extent of 92–93 per cent.

Alternatively such streams may be treated with sulphur dioxide when this is available from an adjacent source.

Since most of the recovered sulphur is ultimately used to make sulphuric acid, it might seem that an economically attractive alternative would be to make sulphuric acid directly from the hydrogen sulphide stream. The main drawback is that it is economically worthwhile to produce and consume the acid only where the hydrogen sulphide stream is available, since the economics of transportation would always favour sulphur. From a process point of view, the combustion of hydrogen sulphide involves the complication that water is produced as well as sulphur dioxide.

One method of treating the wet sulphur dioxide gases involves drying them before they pass to the converters. The gases from the combustion may be partly cooled in a waste heat boiler, but the outlet temperature must be high enough to prevent the condensation of acid in the boiler. The gases are next cooled in a cooling and scrubbing section, which eliminates any dust present. Cooling the wet gas causes the formation of a substantial volume of acid mist from the water and sulphur trioxide present. The acid mist is removed in an electrostatic precipitator. Air is then introduced and the gases pass to a drying tower, where the drying agent is commonly 93 per cent sulphuric acid. The conversion-absorption system is largely on the same lines as for a conventional sulphur-burning contact plant. The essential difference is that the dried gases must be raised to conversion temperature, before entering the first stage of the converter, by heat exchange with gases from the later stages of the converter.

An alternative approach to plants burning hydrogen sulphide permits the elimination of much of the purification section. The gases from the combustion furnace are partly cooled in a waste heat boiler. The temperature of the gases leaving the boiler is adjusted, so that when they are mixed with sufficient air to dilute the sulphur dioxide concentration to 7.5 per cent, the resultant mixture is still hot enough for the preheating, available from the reaction in the first layer of converter catalyst, to raise its temperature to about 438 °C. The gases then pass through the converter in the normal manner, and proceed directly to an absorber at about 438 °C, where the sulphur trioxide is absorbed in 98 per cent sulphuric acid. Sulphuric acid mist is formed in the gas stream, so that gases leaving the absorber are treated in an appropriate type of mist scrubber. This type of plant cannot produce an acid of concentration greater than 93–94 per cent, in view of the moisture content of the gas stream.

The direct production of sulphuric acid from hydrogen sulphide is in commercial operation, but on a relatively modest scale.

The appearance of processes to treat dilute sulphur-containing streams is primarily aimed at the avoidance of pollution. Likewise there has been a whole series of developments aimed at the reduction of sulphur dioxide emissions from a variety of acid plants. Such processes have no specific association with petroleum sources. They involve multiple catalyst chambers, heat exchangers and sometimes duplicate absorption towers. They are not unlike ordinary contact sulphuric acid plants, operated on a more intensive basis. This represents a means of reducing the sulphur dioxide content of stack effluents, from a traditional 2 000 parts per million, to meet statutory requirements limiting the content to perhaps 100 parts per million. This is one of the many manifestations of an increasing concern for the environment.

Part 3

The Industrial Pattern

Chapter 16

Relationship with the Petroleum Industry

In some respects the proposition that petroleum chemicals have a close relationship with the petroleum industry might seem simple and self-evident. The links have, however, varied significantly with time and also with geographical location. The precise impact that one industry has upon the other is frequently obvious enough, but may also be subtle and indirect.

It is appropriate to make a brief, and necessarily very superficial, review of the situation, since the next few years will almost certainly affect this relationship significantly.

The beginnings of petroleum chemicals arose, as was stated much earlier, from the availability of olefinic gas from the refinery processes of thermal cracking and thermal reforming, in the first quarter of this century.

In the U.S.A. to this day 10–15 per cent of the ethylene feedstock and 80–85 per cent of the propylene product comes straight from refineries. The bulk of the U.S. ethylene feedstock is derived from natural gas, which may be regarded as another facet of the petroleum industry.

In much of the rest of the world the production of lower olefins arises from the cracking of liquid feedstocks, and particularly naphtha. These feedstocks, too, emerge from the refinery, but here the link is not quite so close. Products like naphtha and gas oil are articles of general commerce, readily transportable and available in most parts of the world from a variety of sources. Such products used as feedstock lend themselves much more readily to 'arm's length' deals between companies than buying a propylene stream across the refinery fence. Even so the 'ethylene grid' of the U.S. Gulf Coast puts ethylene into virtually the same commodity status.

The main olefinic products, required as raw materials for the petroleum chemical industry, can all be produced in one operation by the cracking of a liquid hydrocarbon feed. All the chemical producer seems to need, therefore, is a supplier for his liquid hydrocarbon, and a lot of capital, for the construction of the cracker itself and for the many associated units required to use the various chemical raw materials to maximum advantage. He will find, however, that he will also produce a considerable volume of fuel gas, a small amount of fuel oil and a really substantial volume of a highly unsaturated, highly aromatic gasoline fraction. The fuel oil and fuel gas can probably be accommodated by the considerable energy needs of the complex but the aromatic gasoline is an important economic factor. It may be economically favourable to extract at least the benzene from this stream, but the remaining gasoline has to be either down-graded to fuel, or hydrogenated, and thereby up-graded, for sale as gasoline. Few chemical producers have found it worthwhile to enter the gasoline market on this basis (though I.C.I. have done so) and one obvious possibility is a reciprocal deal with the petroleum company who provided the original feedstock.

It is apparent from this that, although the possibility exists for a chemical producer to operate independently of the oil industry, other than for the purchase of his feedstock, in practice the links tend to remain close. A great deal of petroleum chemical production is carried out by associate companies of the petroleum industry, and many chemical companies forge formal links with one of the major oil interests.

In the field of aromatics, which has greatly developed in recent years, the process used for their production from petroleum raw materials is largely catalytic reforming. This process is virtually the same as the operation used to make a high octane gasoline component. There may be a specific selection of feedstocks to this process to give a better balance of production for chemical purposes, and, of course, the separation into pure chemicals is entirely divorced from petroleum practice. Nevertheless, such a process can be, and frequently is, carried out within the petroleum refinery, even if the ultimate products are destined as chemical raw materials.

More obviously linked to refinery practice is the recovery of sulphur from sour crudes or sour natural gas. Other essentially refinery products, even if they find use in the chemical field, are naphthenic acids, cresylic acids and naphthasulphonates made from oil.

The position that emerges is that whilst all petroleum chemicals are, by definition, made from petroleum, some are by nature linked very closely with refinery operation; but some may readily be produced in a wholly independent complex which need not necessarily be close to the refinery supplying feedstock.

What is perhaps not so obvious is the impact which the activity of the petroleum industry has on chemical production.

Refineries exist not to make chemical feedstocks but to make petroleum products. In the U.S.A. the pace-setter for the petroleum industry has been the gasoline market. In Europe, historically at least, fuel oil demand has been the factor that has determined the refinery crude runs. The difference arises from a multitude of reasons, but the two main ones are that the U.S.A. is a highly motorized community in a large country consuming vast quantities of gasoline, and that a substantial share of the equally vast U.S. fuel market is taken by the traditionally abundant supply of natural gas.

This difference has meant that in the past the European gasoline fraction was partly available for purposes other than running cars. It was, in fact, put to chemical use with such success that the supplies of naphtha for chemical synthesis have at times been subject to strain. For the future, the pattern of naphtha availability for chemical synthesis in Europe must depend upon the relative growth of the fuel oil and gasoline markets, and the types of crude processed.

The time when a petroleum refinery was largely limited to a simple separation by distillation is decades away. The increasingly varied and stringent demands for petroleum products, and the need to avoid the accumulation of some unwanted fraction, has ensured that a refinery needs to be a highly complex and sophisticated group of plant units.

An important element is the changing nature of the petroleum markets, an immediate example being the current furore over the use of lead additives in gasoline. Much nonsense has been written on this subject, but the core of the argument is that the undesirable exhaust emissions from cars can best be minimized by the use of a catalytic after-burner in the exhaust system (to complete the combustion of exhaust gases) which can only operate in the absence of lead in the gasoline. If this argument is given legislative effect, as seems possible in some areas, a gasoline of satisfactory quality has to be produced without the upgrading traditionally provided by the lead additives.

This need immediately alters the valuation of a number of petroleum products of interest to the chemical industry. The major aromatics, up to a certain limit, represent an obvious way of improving gasoline octane rating, so their value would be enhanced. Alkylate gasoline is also a high octane component and would become in even greater demand. This would consume quantities of the lower olefins from ethylene to amylenes (but more particularly the C_3 and C_4 olefins) in addition to isobutane. The isobutane becomes available in very large quantities from the hydrocracking process. The lower olefins are present in the catalytic cracker gas streams. In Europe alkylate gasoline is not yet important, so that most olefins used for this purpose are derived from U.S. refinery catalytic crackers. Variations in catalytic cracker techniques, which have the effect of increasing the gasoline yield of the process, are likely to affect the yields of the various gas streams.

In the U.S.A. the problems of a switch to the production of an adequate quality of gasoline without additions of lead are enormous, and the costs run into many billions of dollars (billion used here in the U.S. sense of 10^9). In Europe the raw materials (notably isobutane) are largely missing and so are many of the necessary equipment items (such as gasoline alkylation units). The disadvantages of producing a gasoline of marginally reduced efficiency would be greater in Europe where most petroleum is imported. Fortunately the polluting effects are less evident in Europe than the U.S.A. so that the solutions will probably be less drastic. Even if major developments are limited to the U.S.A., the establishment of new values for aromatics and olefins there will have a significant impact on the European scene, since this is a very international industry.

The whole subject of the relationship of petroleum products and petrochemical feedstocks is extremely complex, and a short account can only touch upon some of the major issues involved. These comments have concentrated on the processing implications of this relationship. Clearly any major political issue, which may affect the supply or price of petroleum, will also have an indirect impact on any industry consuming its products.

Chapter 17

Industrial Influence of Petroleum Chemicals

Some indication of the range of activities directly associated with the chemical industry is given by the classification of official statistics. The statistics of the Department of Trade and Industry in the United Kingdom, for example, quote a category described as Chemical and Allied Industries, which includes the following: general chemicals, plastics, coal tar products, dyes and dyestuffs, paints and varnishes, drugs and pharmaceuticals, toilet preparations, fertilizers, disinfectants, polishes, soap, candles and glycerine and explosives.

Many other process industries are extensive users of chemicals, for example the rubber, paper, textile, food and transportation industries. The complex, and sometimes subtle, relationships that exist between the petroleum and petroleum chemical industries have been outlined in the previous chapter.

There are certain industrial sectors where the influence of petroleum chemicals has been particularly marked. The industrial applications of specific products have been outlined in earlier chapters. This chapter examines the position from the point of view of some of the user industries. A selection of such industries for consideration must, to some extent, be arbitrary, though certain industries select themselves. The industries considered in this chapter will help to explain the rapid expansion of petroleum chemicals, since each has grown remarkably in the past two decades.

Plastics

The world production of plastics was 30 million metric tons in 1970. This total has approximately doubled every five years or so for the past twenty years. Looking ahead, it is predicted that world consumption of plastics may reach 45–50 million tons by 1975 and 90–100 million tons by 1980. These figures reveal the tremendous importance of plastics as an actual and potential consumer of organic chemicals.

The world production of plastics may be summarized as follows (figures in million metric tons):

	1950	*1955*	*1960*	*1965*	*1970*
U.S.A.	1.03	1.76	2.85	5.30	8.48
Japan	0.02	0.11	0.55	1.60	5.12
West Germany	0.11	0.37	0.98	2.00	4.33
Italy	0.015	0.10	0.30	0.91	1.73
Soviet Union	0.07	0.16	0.31	0.80	1.67
France	0.03	0.10	0.34	0.70	1.52
United Kingdom	0.15	0.30	0.57	0.96	1.47
Benelux	0.015	0.04	0.12	0.36	1.10
Other countries	0.06	0.36	0.88	1.87	4.58
	1.50	3.30	6.90	14.50	30.00

U.S. PLASTICS PRODUCTION

	Production in long tons 1970
Thermosetting resins	
Phenolic and cresylic resins	465 000
Urea and melamine resins	278 000
Alkyd resins	269 000
Polyester resins	288 000
Epoxy resins	69 000
Thermoplastics	
Low density polyethylene	1 865 000
High density polyethylene	762 000
Polypropylene	463 000
Vinyl chloride polymers and copolymers	1 399 000
Polyvinyl acetate	180 000
Polyvinyl alcohol	22 000
Other vinyl polymers[1]	75 000
Styrene polymers and copolymer resins	1 267 000
ABS/SAN resins (not included in above)	253 000
Cellulose plastics	63 000
Coumarone-indene resins	141 000
All other resins	496 000
	8 355 000

[1] polyvinyl butyral, polyvinyl formal, polyvinylidene chloride

For the next ten years overall world growth, which has run at about 15 per cent per year, is likely to slow down to a slightly more modest 10–12 per cent per year. Consumption should therefore reach marginally over three times the 1970 total by 1980.

Figures are available in more detail for a number of countries. The estimated breakdown of production for the U.S.A. in 1970 is shown on page 268.

Estimates for 1971 suggest a modest overall increase to about 9 million long tons. For 1980 a total consumption in the U.S.A. of about 23 million long tons has been put forward.

Figures for the United Kingdom have been quoted on the following basis:

	Production 1970	*1971 estimates*
Thermosetting resins		
Alkyd resins	65 000	63 000
Polyesters	40 000	38 000
Aminoplastics	135 000	169 000
Phenolic and cresylic resins	59 000	74 000
Polyurethane foams	33 000	48 000
Epoxy resins	13 000	13 500
Others	25 000	n.a.
Thermoplastics		
Cellulosics	10 000	10 000
Polyethylene (low density)	350 000	}
Polyethylene (high density)	66 000	446 000
Polypropylene	73 000	}
Polyvinyl acetate	42 000	40 000
Vinyl chloride polymers and copolymers	310 000	335 000
Styrene polymers and copolymers	159 000	144 000
Others	83 000	n.a.
	1 463 000	1 580 000

There is a significant net export of plastics from the United Kingdom, amounting to 150 000 tons in 1970 (100 000 tons in 1971), so that consumption is less than production.

There is much variation in the consumption of plastics per head in various areas. Some indications are given in the following table, expressed in kilogram consumption per capita:

	1950	*1960*	*1970*
West Germany	1.8	15.0	55.0
Sweden	2.3	11.0	52.0
Japan	0.2	5.8	40.0
U.S.A.	6.4	13.7	39.0
France	1.0	7.4	30.0
Italy	0.6	5.0	28.0
Great Britain	2.5	9.3	23.0
Soviet Union	0.3	1.6	7.0
World (average)	0.6	2.3	8.0

An indication of some of the industrial applications for plastics as a whole may be given by the following estimates for U.S. consumption in 1970:

Plastics in agriculture	88 000 long tons
Usage in major appliances (e.g. refrigerators, washing machines)	159 000 long tons
Usage in small appliances (e.g. TV, radio)	85 000 long tons
Plastics in building and construction	1 260 000 long tons
Usage in electrical/electronic equipment	564 000 long tons
Plastics in furniture manufacture	329 000 long tons
Usage for housewares	431 000 long tons
Plastics for packaging	1 780 000 long tons
Plastic toys	257 000 long tons
Usage for transportation	508 000 long tons

It is clear that outstanding plastics applications are for packaging, appliances and transportation on the one hand—including the various manifestations of increasing affluence—and the building and construction industries on the other. For the future probably the greatest developments in plastics will lie in the building and construction fields. The tendency to replace traditional building materials such as timber, brick, metals and glass by plastics will continue. The replacement may be slow, and it may proceed only to a modest extent, but the total volume of materials involved is so huge that even a modest level of replacement represents a major market for plastics.

The plastics industry has, to a considerable extent, grown up in conjunction with the development of petroleum chemicals, and the two have many links. The association is perhaps most obvious in the field of polyolefins, which have been the most rapidly growing, and are now the largest segment of the U.S. plastics industry. This production, throughout the world, is almost exclusively based upon olefins of petroleum origin.

The world polyolefin consumption in 1970 was about 8 million long tons or around 27 per cent of total plastics consumption. In 1960 only 14 per cent of plastics consumption was in the form of polyolefins. Looking ahead, polyolefins should consolidate their position as the leading group of plastics products. The world demand for polyolefins should reach 15–16 million tons by 1975 and perhaps 30 million tons by 1980. As a percentage of total plastics this represents only a very modest increase, to about 30–33 per cent in 1980. Of the 1980 estimated total it has been suggested that around 56 per cent would be low density polyethylene, with about 22 per cent each of polypropylene and high density polyethylene.

The styrene resins represent another group in which petroleum chemical interest is paramount. Only a small proportion of the world's benzene is now derived from coal; not only is the ethylene a petroleum derivative, but the alkylation process is typical of this field in conception and technique.

The figures for world production of polystyrene are complicated by the range of copolymers which can be involved. If ABS resins are excluded, the production of polystyrene and the other copolymers approximated to 3 million tons in 1970. This figure excludes the communist countries.

The vinyl resins have traditionally been based mostly on acetylene derived largely from calcium carbide. The advent of oxychlorination techniques, to overcome the problems of hydrogen chloride disposal where vinyl chloride is made from ethylene, has provided a tremendous stimulus to the use of ethylene as raw material. Ethylene had become the dominant raw material for vinyl chloride production by 1970. Some of the remaining acetylene facilities will be using hydrocarbon feedstock for the acetylene, and carbide acetylene will play a very minor part in this field. Polyvinyl chloride differs from polyolefins in one respect, in that it has traditionally been pre-eminent among plastics in the European field, and even today the production of polyvinyl chloride in west Europe is approximately equal to that of all polyolefins. In the U.S.A. the position is notably different, with the polyolefins production in total more than double that of vinyl chloride polymers and copolymers.

The world position (excluding the communist countries) is always somewhat confused by different practices in different countries regarding statistics. In 1970 it would seem that nearly 6 million tons of polyvinyl chloride and copolymers were produced.

Vinyl acetate is still mostly based on acetylene. The ethylene-based processes show every sign of becoming dominant in this field as with vinyl chloride. New plants are almost all using one of the ethylene routes, and by 1980 the acetylene route will be largely superseded.

As has already been pointed out, the polyolefins represent the most important group of plastics in scale of production, followed reasonably closely by the vinyl resins, and at a greater distance by the styrene polymers and copolymers. These constitute 'the big three'.

Turning to the general range of plastics and their raw materials, it can be borne in mind that about 90 per cent of the world's organic chemical production today is petroleum-based. Considering plastic groups such as the phenolic, alkyd, urea and melamine resins, the major raw materials include phenol, formaldehyde, glycerine, urea, phthalic anhydride and melamine. Of these, phenol, formaldehyde and urea may be regarded as essentially standard petroleum-based chemicals. Glycerine is still derived mostly from fatty acids outside the U.S.A., where synthetic glycerine is of major significance. Melamine has traditionally been made indirectly from carbide. It now seems established that its future development will lie with the low-pressure routes from urea. Phthalic anhydride is still a major market for coal-tar naphthalene, but the encroachments in this field of o-xylene are now very substantial.

To summarize the position one may refer to the estimate that by 1985 about 98 per cent of the world's organic chemicals will be made from petroleum. It follows that organic polymers will rely essentially on petroleum for their origin.

A further group of plastics is exemplified by the cellulosics, where the fundamental raw material is of natural origin. In such a case synthetic chemicals make their contribution as processing products. It may be expected that petroleum will contribute its normal share of such products.

Coumarone-indene resins are low-priced polymers obtained as secondary materials from the coke and petroleum industries. Their main field is in the production of cheap 'asphalt' floor tiles. Other plastics (notably vinyls) are tending to encroach upon the market for floor coverings. Coumarone-indene resins hold a slowly declining tonnage in an expanding market.

Notable among the groups dismissed as miscellaneous, in the U.S. statistics for plastics production, are methyl methacrylate and nylon. Methyl methacrylate is essentially a derivative of acetone and hydrogen cyanide. Usage in the U.S.A. grew from 56 000 long tons in 1960 to 125 000 tons in 1966. Recent production figures were 214 000 long tons in 1970, and this is forecast to rise to nearly 340 000 tons in 1974.

Methyl methacrylate polymers are commonly encountered as clear sheets for such applications as aircraft windows. Their main shortcoming, which prevents their application to car windscreens, for example, is a relatively poor resistance to scratching and abrasion.

Nylon is obviously of prime concern to the synthetic fibre industry, but the plastics usage has risen in the U.S.A. from about 15 000 long tons in 1961 to 46 000 tons in 1970. This material may conveniently be used in small bearings and hinges, where advantage may be taken of its self-lubricating properties. Nylon is a generic term, and most of the polyamide resins listed under this heading are basically of the nylon 6/6 type (based on adipic acid and hexamethylene diamine). Some nylon 6 (based on caprolactam) is also used in resin or film form. There is a whole range of other nylons (designated *inter alia* 6/12, 6/10, 11, 12) which are high-priced specialities, with a very small market at this stage. Although nylon resins have taken quite a long time to develop, it is now expected that growth should accelerate, and a total of over 150 000 long tons in the U.S.A. has been estimated as the 1980 usage. Worldwide the market for the nylon resins exceeded 150 000 long tons in 1970.

Amongst the newer plastics which may be said to have 'arrived' are the polyester resins (by which is normally understood unsaturated polyester resins), epoxy resins, polyurethanes and polycarbonates.

The figures for polyester and epoxy resins usage in the U.S.A. have developed as follows (in '000 long tons):

	1947	*1954*	*1960*	*1966*	*1970*
Polyester resins	2	13	85	182	288
Epoxy resins	—	10	29	54	69

In the United Kingdom, polyester resins, which were used on a scale of a few hundred tons in 1954, were produced to the extent of 40 000 tons in 1970. Epoxy resin growth has been slower than that of the polyester resins in the United Kingdom, as in the U.S.A., and the level of production was about 13 000 long tons in 1970.

Polyester resins are largely used in conjunction with glass fibre, to produce the 'reinforced plastics' used in the manufacture of boats and other transport equipment. Epoxy resins are also used in conjunction with glass fibre, but in this case the end product is more commonly in the form of specialized engineering 'tools', for the mass production of pressed steel units for car bodies and aircraft parts. Epoxy resins are also of interest in the development of specialized surface coatings with an exceptional resistance to corrosion and impact, and for the protection of electronic assemblies.

Polyurethanes have exhibited one of the most explosive developments of the past decade. Originating in Germany, they have undergone spectacular commercial development in the U.S.A., illustrated by the following production figures (in '000 long tons):

	Flexible foams	*Rigid foams*	*Non-cellular*	*Total*
1955	0.5	0.5	0.5	1.5
1960	42.5	5.4	9.8	57.7
1965	111.6	31.7	31.7	175.0
1970	300.5	100.0	47.4	449.9
1975 (est.)	503.0	287.0	90.0	880.0

Foams, both flexible and rigid, represent the largest and fastest growing outlet for polyurethanes, as the figures show. Amongst the non-cellular applications, one of the most notable is for elastomers. This took nearly 23 000 tons of polyurethanes in 1970, or about half the non-cellular total. It should hold at least this percentage, according to the forecast of almost 50 000 tons for urethane elastomers in the U.S.A. for 1975, out of the total non-cellular market of 90 000 long tons forecast for that same year. Urethane elastomers find application in spandex stretch fabrics, such as Lycra, used for ladies' foundation garments.

The development of the flexible foams for cushioning has been, to some extent, at the expense of foamed rubber latex. The polyurethane foams have advantages in that they are lighter in weight and easier to process than their rubber latex counterparts.

The rigid polyurethane foams were slower to gain acceptance than the flexible foams but are widely used as a lightweight but durable insulating material in construction work and appliances.

The polyurethane surface coatings are also of significance. It is dangerous to refer in book form to such transitory matters as fashion, but the 'wet look' so much in demand at the beginning of the 1970's was essentially based on polyurethane finishes. In the form of more conventional materials, polyurethanes can provide surface coatings of high gloss and durability, with an exceptional resistance to impact.

The worldwide production of polyurethanes was about 1 million tons in 1970 and the growth is such that the figure could reach 2 million tons by 1975.

To the list of the established plastics one should now tentatively add polycarbonates. These products have made a slower start than was predicted a few years ago. In 1962 the U.S. consumption was about 2 000 tons. It was over 9 000 tons in 1966 and reached about 16 000 tons in 1970. An important characteristic of polycarbonates is that, although they are thermoplastics, they have a resistance to heat which compares favourably with that of many thermosetting resins. The electrical properties are favourable, and they have exceptional transparency. This represents one of the group of 'engineering plastics' which may be expected to have a buoyant future in various aspects of the construction field.

One becomes a little wary of buoyant forecasts for polymers, more particularly for polycarbonates which have already exhibited disappointing growth in the face of great expectation. One should, nevertheless, note that a recent U.S. forecast put the 1980 usage of polycarbonates there at nearly 70 000 long tons. The consumption of polycarbonates in Europe in 1970 was about 15 000 tons, about 67 per cent of which was in west Germany, where Bayer, the pioneers of this material, are the sole European producers. The European pattern of usage has been indicated:

Automotive and mechanical engineering	33%
Lighting	24%
Aeronautical and military applications	23%
Electrical and electronic uses	12%
Miscellaneous (inc. tableware)	8%

The world consumption of polycarbonates in 1970 was around 50 000 tons.

Another group of resins, also in a relatively modest stage of commercial development, and also included in the range of 'engineering plastics', is represented by polyformaldehyde or the polyacetals. These products may be either straight formaldehyde polymers (e.g. Delrin) or copolymers in which a small proportion of a comonomer such as ethylene oxide is incorporated (e.g. Celcon). These are products of considerable strength, which may, in certain applications, replace metals such as zinc and aluminium. The consumption in the U.S.A. was 26 000 long tons in 1966, and this proved to be a peak figure. It dropped to around 16 000 tons in 1967, and has since steadily increased to almost 28 000 long tons in 1670. Here again, one is faced with buoyant forecasts for the future. A figure estimated for U.S. consumption of polyacetals in 1980 was 90 000 long tons. The world consumption of polyacetals in 1970 was about 70 000 long tons, of which about 19 000 tons was in Europe.

The breakdown of the U.S. demand in 1970 has been quoted;

Automotive	23%
Mechanical engineering	20%
Household appliances and consumer goods	20%
Ironmongery	14%
Electrical applications	11%
Magnetic tape supports	11%
Miscellaneous	1%

The newer groups of plastics which have been outlined have enjoyed a relatively rapid growth in usage, and are expected to expand much further in the next decade. They all lean very heavily on petroleum chemicals, as is indicated by the brief list of typical raw materials given opposite.

This is a very forward-looking segment of industry and it is always necessary to take note of newer polymers, some of which may be poised for substantial development. Considering products under this heading, one should draw attention to the fact that the market place is littered with 'failed polymers'. Ten years ago the market was agog for the next generation of products to develop along the lines of the polyolefins. It is now more soberly believed that it is unlikely for new products to find their way into the top league, and the most that is

Polyester resins	Ethylene glycol, propylene glycol, maleic anhydride.
Epoxy resins	Epichlorohydrin, diphenylolpropane.
Nylon	Adipic acid, hexamethylene diamine, caprolactam.
Polyurethane resins	Toluene diisocyanate, polyethers (e.g. from propylene oxide and glycerine).
Polycarbonates	Diphenylolpropane, phosgene, diphenyl carbonate.
Polyformaldehyde	Formaldehyde.

expected of polymers now put forward for assessment is that they will find their place among the 'also-rans' of the plastics industry. The level of price and performance of the major plastics is such that there does not seem room for additional general purpose products. A new polymer is therefore likely to be doomed to being a more-or-less speciality product. The development of such materials continues apace.

Polyphenylene oxide products are polymers of a synthetic cresol, 2,6-xylenol. They may be used over an exceptionally wide temperature range from −170 to over 160 °C. They have good properties for applications in the construction and engineering field as a replacement for metals. The products made their commercial debut in the winter of 1964–65. The original homopolymer has made rather slow progress, but the prospects seem to have improved by the introduction of polystyrene-modified material (e.g. Noryl). By 1970 world markets were said to be 3 000 tons for polyphenylene oxide as such, and 7 000 tons for the polystyrene-modified material. It has been suggested that the market for the modified polymer might approach 30 000 long tons by 1975.

The parylenes are essentially polymers of p-xylene, but the crucial step is the intermediate production of a cyclic dimer compound, di-p-xylylene. This dimer can be cleaved to the monomer p-xylylene, which, when adsorbed on a substrate instantly polymerizes in a continuous film. Whilst this ability to form accurately uniform films on surfaces of various shapes is of interest in a number of applications, the particular aspect that excites many engineers is that a thin film of parylene deposited on condenser tubes promotes dropwise rather than filmwise condensation of water. This has an immediate impact in minimizing the insulating effect of a water film on condenser tubes and therefore increases heat transfer. This may well be of significance in improving the economics of such processes as the desalting of sea water. Little has been heard of these polymers in the past year or two.

Polysulphones are complex aromatic polymers in which the distinctive feature of the basic unit is a diphenylene sulphone group. This is another field in which diphenylolpropane has potential application. The structure of the polymer has been quoted as

$$\left[-C_6H_4-C(CH_3)_2-C_6H_4-O-C_6H_4-SO_2-C_6H_4-O- \right]_{50-80}$$

The combination of properties includes toughness, rigidity, oxidation resistance and, relatively unusual in the plastics field, non-flammability. These products may be used over a wide temperature range, and the electrical properties are good. It is still a relatively high-priced material, but will find applications in the engineering field. A plant of 4 500 tons per annum capacity is in operation in the U.S.A. The world consumption of polysulphones in 1970 has been put at about 3 000 tons.

The TPX resins are polymers of 4-methylpentene-1, a dimer of propylene. These products have been developed in the United Kingdom by I.C.I. from a monomer prepared by BP.

They are clear glass-like polymers with excellent electrical properties and an exceptionally low specific gravity. A small commercial unit of capacity about 2 000 tons per year is operated by I.C.I. at Wilton. Progress in the development of the homopolymers has been slow. Copolymers with other olefins show promise.

Mention should perhaps be made of the fluorocarbon plastics. These have been produced on a relatively modest scale for many years. The source of most of this production is the hydrofluorination of chloroform to give dichlorodifluoromethane. The dehydrochlorination of this gives tetrafluoroethylene. This is the monomer for perhaps 90 per cent of all fluorocarbon plastics. The world consumption of these in 1970 was around 13 000 tons, of which 8 000 tons was in the U.S.A.

Two new polymers of which considerable expectations are held in some quarters are polybutylene terephthalate (from terephthalic acid and 1,4-butanediol) which had a negligible production in 1970 but has been forecast to reach a 20 000 ton market in the U.S.A. by 1975, and polybutene-1, a Ziegler-type polymer of the 1-butylene isomer. Polybutylene terephthalate is under development as an engineering plastic by Celanese and General Electric Co. in the U.S.A. Hüls have operated a 1 200 tons per year pilot plant for polybutene-1 and were scheduled to complete a 12 000 tons per year commercial unit early in 1972. The homopolymer was particularly suitable for pipe manufacture, and copolymers with other olefins are expected to extend the range of applications.

In thinking of the future of plastics, one cannot ignore the implications of techniques in current use. Reference has been made (in the cases of polypropylene and polybutadiene, for example) to the use of Ziegler-type catalysts, comprising essentially aluminium alkyls, in conjunction with titanium halides, to provide polymers of a stereospecific nature. By this is meant that from an appropriate monomer we can prepare a polymer of predetermined regular molecular configuration. Knowledge of the influence of molecular configuration on the physical and chemical properties of polymers is not yet at a stage where we can fully appreciate or develop the implications of this possibility. There is little doubt, however, that further industrial developments of this type are in store.

On a global basis, the growth of the plastics industry has been one of the outstanding industrial achievements of the past decade. The growth rate for plastics has been faster than that of the chemical industry, which itself has been faster than that of the general world economy. The maintenance, in the future, of a growth rate in any way comparable with that of the past decade, depends on the application of plastics to new fields. It has already been indicated that the present usage of plastics is particularly concentrated in such fields as packaging, vehicles, domestic appliances and building. The expansion of the first three of these is clearly sensitive to the constant trend towards higher living standards. The share taken by plastics in each vehicle and appliance is also tending to grow continuously. But, as mentioned earlier in this chapter, it is in the field of building, and the tendency to replace traditional materials by plastics in this whole vast construction industry, where the really big developments may be seen in the longer term.

Whatever the individual prospects for specific resins, there seems little doubt that there will be a further major expansion in the production and use of plastics (and hence of petroleum chemicals) in the next decade. Moreover, it is those products most closely associated with a petroleum derivation which are proving the pacemakers in this field.

Synthetic Fibres

It has become customary to regard the term 'man-made fibres' as including derivatives of natural products (essentially the cellulose-based artificial fibres), as well as the wholly synthetic products, whereas the latter are described as 'synthetic fibres'.

The world production of synthetic fibres in 1970 amounted to over 4.5 million long tons, a figure nearly seven times the output for 1960. The U.S.A. was responsible for about 35 per cent of this production, western Europe (including the United Kingdom) 33 per cent, Japan 22 per cent and other countries 10 per cent.

Nylon holds the largest share of the market, although this share is declining. Immediately after 1945 the nylon fibres represented virtually the whole synthetic fibre production. By 1960 their share had dropped to under 60 per cent of the total, by 1967 the proportion had dropped to 46 per cent, and by 1970 to marginally under 40 per cent. Polyesters have been increasing their share of the total, from 17 per cent in 1960 to 27 per cent in 1967 and almost 34 per cent in 1970. The corresponding figures for acrylic fibres are 15, 19 and 19 per cent. There remains a miscellaneous category, including fibres based on polyolefins, polyvinyl alcohol, polyvinyl chloride, elastomeric materials and fluorine-based products. This has remained largely static, as a proportion of the total, having been 9 per cent in 1960 and 8 per cent in 1970.

Looking to the future, it is to be expected that wholly synthetic fibres will grow to about 8.1 million long tons consumption in 1975. By then the share of nylon and the polyesters will be about equal at 35 per cent each, with 20 per cent for acrylics and 10 per cent for the others.

Production of synthetic fibres in the United Kingdom in 1970 amounted to about 335 000 long tons or something like 7.5 per cent of world production.

The total world apparel fibre market (including wool, cotton and the man-made cellulose fibres as well as the synthetics) was almost 15 million long tons in 1960, had reached almost 18 million in 1966, and topped 21 million long tons in 1970. An estimate of the total for 1975 is about 25.5 million long tons. The share of synthetics rose from 4.5 per cent of the total in 1960 to over 21 per cent in 1970. By 1975 the proportion could be over 31 per cent. Further development is also expected in the use of synthetic fibres to replace 'hard' fibres such as flax, jute and hemp.

There is a good deal of over-capacity for synthetic fibres in many of the industrialized areas of the world at the present time—more particularly in view of the poor showing of 1970. The future growth is likely to be most marked in the communist countries, and, for the longer term, in developing countries, so that over-capacity in the western world may take some time to disappear. It is also apparent that the growth in world demand for synthetic fibres is likely to slow down in future years. The reduction in percentage growth is a normal characteristic of an expanding product, as the base figure (on which the percentage is calculated) grows larger.

The raw materials for the main forms of nylon have been discussed in previous chapters. To an increasing extent this production is based on cyclohexane from petroleum-derived benzene. Phenol is an alternative raw material both for adipic acid (via cyclohexanone) and for caprolactam. The precursor for hexamethylene diamine, adiponitrile, may now be made from acrylonitrile, instead of the more conventional alternatives of adipic acid or butadiene. The route to adiponitrile from furfural (which is not derived from petroleum) has now been abandoned. Toluene is an alternative raw material for caprolactam. The only significant residual influence in this area of non-petroleum raw materials is the use of coal tar benzene. This has been important in Europe. The major switch to petroleum sources for benzene in Europe has eliminated much of this residual effect.

After nylon, the polyesters are the next most important group of synthetic fibres on a world-wide basis. They have always been relatively more important than acrylic fibres in Europe, although the reverse used to be true of the U.S.A. This may reflect the European (or rather, British) origin of the polyester fibres. The essential basis of polyester fibres is the reaction between ethylene glycol and terephthalic acid (which is, for convenience, normally used in the form of the dimethyl ester). As has previously been indicated, ethylene glycol is a major ethylene derivative, via ethylene oxide, and terephthalic acid is normally an oxidation product of p-xylene from a refinery catalytic reformate.

Acrylic fibres consist essentially of polyacrylonitrile, and are therefore almost exclusively of petroleum origin.

The remaining fibres have, so far, been much less important in scale of production. The interest of petroleum in this range of fibres, however, is once again substantial. Polyolefin fibres are almost exclusively based on olefins from catalytic cracking or naphtha pyrolysis, and the vinyl copolymers are largely derived from ethylene dichloride.

Synthetic Rubbers

After the very necessary substitution of natural rubber by synthetic rubber in the latter years of the second world war (with most of the world's rubber plantations in the hands of the Japanese), the scale of production tended to drop. Production of synthetic rubber was again stimulated by the hostilities in Korea in the period 1950–51, and it has remained at a higher and increasing level since. This movement is due partly to a substantial replacement of natural rubber by synthetic in the passenger tyre market, but, more fundamentally, to the fact that the production of natural rubber cannot be expanded rapidly, since it takes about seven years for new plantations to commence production. As a result, the availability of natural rubber during the 1950's was not increasing as rapidly as the total demand for new rubber. Unstable political conditions in some of the rubber-growing areas have also tended to restrict the development of natural rubber.

The production of natural rubber has moved slowly and steadily up from just over 2 million long tons in 1960 to almost 2.9 million long tons in 1970. The production of synthetic rubber (world figures including Czechoslovakia and Poland but excluding other communist countries) during the same period has grown from 1.94 million long tons to 4.87 million. The percentage of the total taken by the synthetic products has therefore risen from 52 to 66 per cent.

In 1970 the consumption of rubber is given as 7.62 million tons.

This is a slightly hybrid figure since it excludes the production of eastern European countries (other than Czechoslovakia and Poland) consumed within this area. The world (excluding communist countries) is forecast to use almost 11 million tons of rubber by 1979, and of this 73 per cent is likely to be synthetic. By 1985 an estimate has put the usage at 15 million tons, 75 per cent synthetic.

Among individual countries the percentage consumption accountable to synthetic material in 1970 was 77 per cent in the U.S.A., 59 per cent in the United Kingdom and about 60 per cent in western Europe as a whole. Other percentages varied from 27 per cent in India to 73 per cent in Canada.

The consumption per head of total rubber in 1970 was highest in the U.S.A. at about 29 lb; in the United Kingdom the figure was rather under 20 lb, whereas in India it was less than 1 lb.

Production of synthetic rubber is a major industry in many countries. The United States has by far the largest scale of production, the total for 1970 being over 2.23 million tons (or, excluding the oil content of SB rubber, about 1.98 million tons). Other major producers in 1970 were Japan with 698 000 tons, France with 316 000 tons, the United Kingdom with 306 000 tons, west Germany with 302 000 tons. Canada and the Netherlands were almost equal with 205 000 tons. Canada was one of the earliest major producers of synthetic rubber, but her production has been almost static since 1964.

The most detailed figures are available from the U.S.A. as the following summary of production statistics indicates (in '000 long tons):

	SB* *rubber*	*Nitrile*	*Butyl*	*Poly-butadiene*	*Other stereo-specific rubbers*	*Others†*	*Total*
1960	1 185	39	99			136	1 459
1965	1 282	60	102	156	66	176	1 842
1970	1 352	68	120	284	186	222	2 232

* Includes oil content.

† Includes stereospecific rubbers in 1960 and in subsequent figures silicone, urethane and polysulphide rubbers. The main element in this total throughout is polychloroprene, rising from about 100 000 tons in 1960 to perhaps 170 000 tons in 1970.

There are no official figures giving a breakdown of United Kingdom production but an approximation for 1970 would be:

SB rubber (solid and latex)	72%
Polybutadiene	14%
Nitrile (rubber and latex)	5%
Butyl	9%
Polychloroprene (rubber and latex)	8%
Others	6%

The capacity for synthetic rubber (excluding the communist countries) is divided approximately as follows:

	1970	*1973* (est.)
SB rubber	60.5%	54.5%
Polybutadiene	16.9%	19.0%
Butyl rubber	6.2%	5.3%
Polychloroprene	5.6%	6.1%
Nitrile rubber	4.5%	4.0%
Polyisoprene	3.7%	6.2%
EP rubbers	2.6%	4.9%

Although it can be seen that there are significant differences between the rates of growth of the different rubbers, the outstanding importance of the SB rubbers will continue, with polybutadiene assuming a second place, well ahead of the other contenders. Butyl and nitrile rubbers are expected to grow relatively slowly in comparison with polyisoprene, EP rubbers and polychloroprene.

The main basis for the rubber industry as a whole will continue to be in tyres and related products, which account for something like 60 per cent of all rubber consumption.

The range of rubbers now available will, in theory, meet almost all the technical requirements for the rubber-consuming industries. Polybutadiene and polyisoprene may be used as a complete or partial replacement for natural rubber in the heavier grades of tyre where SB rubber is not suitable. The growing usage of radial ply tyres has given some new impetus to the use of natural rubber in car tyres, but here polyisoprene is also suitable. The introduction of ethylene–propylene rubbers into the tyre industry in the near future is not now predicted with quite the same confidence as it was a few years ago, but once the technical problems are overcome they should be very economical materials to use in view of their light weight.

The listing of the appropriate raw materials for the synthetic rubbers gives an immediate indication of the overwhelming importance of petroleum chemicals in this field. Apart from the obvious need for chlorine in polychloroprene, there is only the fast declining usage of carbide acetylene (also for making polychloroprene) which does not have a petroleum basis. The raw materials are summarized as follows:

SB rubber	Butadiene, styrene
Polychloroprene	Acetylene, butadiene, chlorine
Nitrile rubber	Butadiene, acrylonitrile
Butyl rubber	Isobutylene
Polybutadiene	Butadiene
Polyisoprene	Isopentene, propylene, isobutylene, formaldehyde
EP rubbers	Ethylene, propylene

Historically the natural rubber market has been notable for its erratic fluctuations, but the introduction of synthetic rubbers on a substantial scale has ensured that the prices of natural rubber have been stabilized and moved generally downwards, reflecting improvements in the planting and operation of natural rubber plantations in the past decade. A possibly significant factor in natural rubber production is the potential use of ethrel (2-chloroethylphosphonic acid) as an anticoagulant. The application of ethrel to rubber trees apparently releases ethylene and causes the trees to bleed and release more fluid. What the long term effect on the trees may be has yet to be established, but in the short term the yield per acre (and presumably per employee) can be increased.

The slow but steady increase in natural rubber production may well continue, but it is clear that the overall market will be dominated by the synthetic material.

Detergents

Detergency, as we normally understand it, is associated with the phenomenon of surface activity on the part of the detergent molecules. Surface activity involves the orientation of detergent molecules in regular patterns, more particularly at surfaces and interfaces, but frequently also in aggregations within the body of the liquid, called micelles. These characteristics enable the detergent solution to exhibit phenomena such as reduced surface tension, foaming, and soil stabilization, which are normally associated with a detergent effect.

The traditional detergent is soap. Soap possesses a typical molecular structure for a surface active agent. It has a long hydrophobic (i.e. water-repelling) hydrocarbon chain, at one end of which is a hydrophilic (i.e. water-attracting) group. As a household detergent and industrial wetting agent, soap possesses two important disadvantages: first, it forms insoluble calcium salts which are precipitated from hard water, and secondly, it is unstable in acid solutions.

The search to find products which would provide the detergent effect, without these disadvantages, has proceeded over a long period. The first commercial developments of major significance were in the 1930's, when primary alkyl sulphates became available, using as a basis primary alcohols obtained by the reduction of fatty acids. During the second world war alkyl sulphonates were used to a substantial extent in Germany. After 1945 there was some development in Europe of secondary alkyl sulphates based on olefins from wax cracking. The major development proved to be that of alkyl benzene sulphonates. Until quite recently propylene tetramer was the favoured form of alkyl group for this production. The increasing use of such products gave rise to problems in the treatment of industrial and domestic effluents, as the normal sewage treatment processes using biological degradation of the detergent material coupled with precipitation of soap by vast volumes of water present, did not succeed in disposing of the surface active material present. The tetrapropylene benzene sulphonates were resistant to biological degradation. Modified alkyl benzene sulphonates are now used in which every effort is made to provide a straight chain alkyl group, which has been found to be more 'biodegradable'. The processes involved include aluminium chloride alkylation using the monochloride derivatives of n-paraffins, and conventional hydrogen fluoride alkylation using straight chain olefins made from urea-treated wax, or from n-paraffin dehydrogenation.

These products are almost completely, but rather slowly, degraded by the biological processes of sewage treatment. The search is for more rapidly degraded products. Amongst the contenders for the next generation of major detergents are primary alkyl sulphates from detergent alcohols of the C_{12}–C_{18} range made by the Oxo process; alkenyl sulphonates of the C_{15}–C_{20} range made by a mild sulphonation of straight chain alpha olefins; products from the 'ethoxylation' of primary or secondary long chain alcohols with 1 to 3 molecules of ethylene oxide per molecule of alcohol, including the ether sulphates made by the sulphation of the 'ethoxylate'; and a new form of alkyl sulphonate in which the alkyl component is based on n-paraffins.

These products, with the exception of the 'ethoxylate' group, all ionize in solution and are described as anionic, since the long chain anion is deemed to impart the specific properties of surface activity. The other type of ionizing detergent is described as cationic. Cationic detergents (e.g. cetylpyridinium chloride), in which the cation is the hydrophobic group, have never developed beyond the level of relatively small-scale industrial and laboratory specialities.

There is another category of detergent which has in recent years found a substantial market; this is the nonionic detergent, which does not ionize in solution. The main type of nonionic detergent is a condensation product of a number of molecules of ethylene oxide with a hydroxy compound incorporating a long hydrophobic alkyl chain. The condensation adds a number of ethylene ether groups, which are hydrophilic, to balance the activity of the molecule.

As the hydroxy compounds, the traditional products were alkyl phenols, which could be condensed with 8–15 molecules of ethylene oxide per molecule of alkyl phenol, to give a range of commercial products. Adjustments to the length of the hydrophobic alkyl chain, or the extent of addition of ethylene oxide, could control the water-solubility and other characteristics of the finished product.

Here again, the question of biodegradability has reared its head. The alkyl phenol nonionic condensates may be slowly degraded, but the corresponding long chain alcohol derivatives have similar surface active properties and are degraded much more easily. The nonionic ethoxylates used commercially include typically C_{16}–C_{18} alcohols with 10–25 molecules of ethylene oxide per molecule of alcohol (these are good low-foaming detergents), or C_{10}–C_{15} alcohols with 6–8 molecules of ethylene oxide per molecule of alcohol (these make good wetting agents and liquid detergents, acting as direct replacements of the most common alkyl phenol derivatives). To complete this picture one should add the condensation of C_{10}–C_{18} alcohols with 3–10 molecules of ethylene oxide per molecule of alcohol. Such ethoxylates are commonly sulphated and neutralized to give anionic ether sulphate detergents which are used in a variety of general purpose cleaning preparations.

It has been suggested that in 1971 the proportions of the various types of synthetic detergents across the world would be anionic products 62 per cent, nonionics 29 per cent and cationics 7 per cent. The proportion of cationics will be small everywhere but the proportion of the total taken by the nonionics varies widely. Individual countries may be quoted as follows:

	% of total synthetic detergents which are nonionic
France	22
Germany	32
U.S.A.	28
Japan	40

In most countries it can be seen that the main basis of synthetic detergent production is represented by anionic detergents, notably the alkyl benzene sulphonates. Whilst the need for these products to be 'linear' (i.e. to be based on a straight alkyl chain) is by no means universal, legislation, which has the effect of eliminating the use of the branched chain products, is becoming more common, and even where the technical need for a high level of biodegradability cannot be established, it is increasingly fashionable to insist on the straight chain products. In a sense this is a pity, since of all the anionic detergents it is generally agreed that tetrapropylene benzene sulphonate has the best properties in every respect except that of biogradability. Detergents and emulsifying agents based on tetrapropylene benzene were still produced for specialized industrial purposes up to the early 1970's but are now declining rapidly, and are likely to be eliminated from the western European scene before 1975.

The world capacity for the linear alkyl benzenes was nearly 800 000 tons per year by the end of 1971. These products are more expensive to produce than tetrapropylene benzene, and in the earlier stages presented some processing problems. It is true to say that no one is yet satisfied that these products represent the end of the search for biodegradable detergents, since the degradation occurs relatively slowly, and is not at all satisfactory in the special conditions which call for treatment in anaerobic conditions (i.e. in the absence of air—

this can be required in areas serviced by septic tanks rather than sewage systems).

The demand for linear alkyl benzenes in the U.S.A. is rising slowly. The 1971 figure was slightly under 250 000 long tons, and the forecast for 1975 is scarcely 280 000 long tons.

So far as the nonionic surface active agents are concerned the 1970 production was around 885 000 long tons (of which slightly over half was consumed in the U.S.A.). It has been suggested that by 1975 the world level of production of nonionics will reach 1 475 000 long tons. Of this total over 1.2 million long tons will be produced in the U.S.A. and Europe. This production will be based to the extent of 355 000 long tons on alkyl phenols, and 580 000 tons on long-chain alcohols. There are many other types of nonionic surface active agents, including amides and esters of carboxylic acids.

The world production of cationic surface active agents was said to be 275 000 long tons in 1970. These seem to be developing especially rapidly in the countries of eastern Europe, since it has been indicated that they expect to expand from a 3.3 per cent share of the world's 1970 production to a 9 per cent share of the 590 000 long tons forecast as the world's production in 1980.

Comparing figures for soap and synthetic detergents, it is necessary to note that these quantities normally relate to products as formulated for sale, rather than as 100 per cent surface active agents. Whilst this may be confusing and lead to inaccuracy, it incorporates a certain rough justice in that such products are provided for sale on a competitive performance basis.

UNITED KINGDOM (figures in '000 long tons)

	Soap deliveries	*Synthetic detergent sales*	*Total*	*Synthetic detergent %*
1964	383.0	337.3	720.3	46.8
1966	338.8	365.4	704.2	51.8
1967	320.0	387.1	707.1	54.8
1968	311.3	413.9	725.2	56.3
1969	275.3	483.9	759.2	63.7
1970	261.7	498.0	759.7	65.7

The inclusion of exports slightly obscures the picture of gradually increasing consumption in the United Kingdom. In 1951 only 15 per cent of the total consumption of 667 400 tons was synthetic. The proportion of synthetic material rose to 33.4 per cent in 1955 and to 40 per cent in 1960.

The rate at which synthetic detergents have replaced soap has been relatively slow in the United Kingdom. In Japan synthetics had taken 68 per cent of the market in 1965 and 82 per cent in 1970.

In the U.S.A. the share of synthetic detergents was only 4 per cent in 1945, but this had risen to over 30 per cent in 1950 and to above 80 per cent by 1964. Since then there has been little change in the U.S.A., with the proportion of total detergent materials attributable to synthetics creeping up to near 85 per cent. The total consumption of soap, synthetics and other allied products reached 4 million long tons in 1968, and this total is rising slowly.

The position in Europe has been assessed as follows (in '000 long tons):

WESTERN EUROPE

	Soap consumption	*Synthetic detergent consumption*	*Synthetic %*
1960	1 360	1 200	47
1965	1 240	2 160	63.5
1970	950	3 190	77
1975 (est.)	895	3 990	81.5
1980 (est.)	760	4 920	87

EASTERN EUROPE

	Soap consumption	*Synthetic detergent consumption*	*Synthetic %*
1960	1 220	290	19
1965	1 460	570	28
1970	1 350	1 215	47.5
1975 (est.)	980	2 200	69
1980 (est.)	865	3 300	80

For the world as a whole the pattern has been presented as follows (in '000 metric tons) for 1969:

		% of total
Total world production	16 130	100
Production of soap	6 440	40
Production of synthetic detergents	8 050	50
Production of scouring agents	560	3.4
Production of auxiliary washing agents	1 080	6.6

The figures on a similar basis for 1960 gave soap 63 per cent of the total and synthetic detergents only 31 per cent.

Once again it should perhaps be emphasized that the presence of auxiliary washing products, and international differences in washing practices, demand caution in interpreting these figures on a comparative basis, but the overall trends are clear enough.

It has sometimes been indicated that the total soap and detergent consumption will give a rough indication of the standard of living, but the comparisons indicated by the figures below are, at best, approximate:

	Total annual per capita usage of soap and detergent (lb)	*Synthetic detergent %*
U.S.A.	44	84
United Kingdom	28	65
Western Europe	28	77
Soviet Union	13.5	40
South America	9.5	—
Asia	3	—
India	2.5	4
China	0.7	2

In the more industrially advanced countries the main residual soap market is for toilet soap in bar form. There has not so far been any acceptable synthetic detergent alternative to toilet soap. In the United Kingdom the switch to synthetic detergents has, for reasons never clearly explained, not proceeded quite so far as in other industrialized countries. We are almost unique in retaining a certain allegiance to soap-based washing powders. This may be due to inherent conservatism, or possibly a recognition that soap is not without its merits as a means of general purpose washing.

For general washing purposes, both soap and synthetic detergents are commonly used as formulated products, spray-dried into the form of small hollow particles known as 'beads'. The main products used in conjunction with synthetic detergents in such formulations are termed 'builders'. These may include sodium salts such as silicates, carbonate and sulphate, together with foam boosters, anti-deposition agents (such as carboxymethyl cellulose), optical brighteners and enzyme stain-removal agents. The most important builder is a polyphosphate (notably trisodium polyphosphate). These polyphosphates are currently under attack by virtue of the undesirable phenomenon known as eutrophication. Accumulations of phosphates in stagnant waters and lakes exert an effect, typical of a fertilizer, in stimulating growth of plant life such as algae. This development, apart from being unsightly, minimizes the oxygen available in the water for maintaining other forms of aquatic life.

As with all environmental arguments, there are many inconclusive features. It should first be emphasized that the problem is limited to areas where much effluent flows into lakes (e.g. the U.S.A., Sweden, Switzerland). Phosphates are present in these waters from many other sources than detergent products. Nevertheless, since detergents are an identifiable source of part of this type of pollution, consideration is being given to the use of other builders.

This study has drawn attention at least to the fact that these polyphosphates are extremely valuable detergent additives, and it is very difficult to find a replacement which will provide all their attributes. The prospects of nitrilotriacetic acid were widely canvassed in the late 1960's, but these received a jolt in 1970 with the dropping of their use in the U.S.A. on account of an alleged toxicity hazard. More recently sodium citrate (normally made by a fermentation process from molasses), and disodium oxydiacetate (which may be made from diethylene glycol or from chloroacetic acid) have been recommended for this purpose, but they have not yet established themselves. The same comment applies to the benzopolycarboxylic acids.

An incidental factor of interest in this field, which is not without a certain irony, is that the development of synthetic detergents, in addition to its effect on the consumption of soap, also undermined the established basis for the production of glycerine. Petroleum chemicals made amends, however, by developing an alternative basis for glycerine supply, in the form of processes for the production of synthetic glycerine. In some respects this has reacted to the advantage of glycerine, since the economic basis of a material of wholly by-product origin is always a little suspect. Production and usage of glycerine are today higher than ever, partly assisted by the stability provided by the availability of a synthetic product.

Amongst detergent products, where soap has held undisputed sway for centuries, the synthetic detergents have made a very note-worthy impact within the brief space of approximately 25 years' active trading (and much less in many parts of the world). This development may be claimed as another industrial sphere in which petroleum chemicals have exerted an impressive influence.

Other Industries

It is a platitude that the chemical industry is its own best customer. The major chemical companies are all substantial users of chemicals (including their own chemical raw materials) as well as producers. It is not, however, possible to develop a satisfactory statistical basis to cover such a wide range of internal transactions.

Any industry deeply concerned in the consumption of organic chemicals must (in the light of the statistical background) be a major consumer of chemicals from petroleum. Such a definition clearly applies to such classifications as dyestuffs, drugs and pharmaceuticals and toilet preparations.

The field of inorganic chemicals from petroleum is much more confined, as has been pointed out. Petroleum-based products touch upon many industries in the form of sulphur or sulphuric acid. Explosives frequently involve the use of nitric acid, a primary derivative of ammonia.

To study in detail the whole range of industries with an interest in the use of chemicals from petroleum would tend to become repe-titious, and in some cases the link of the final product with its initial petroleum origin might become tenuous. In a very broad picture, let us indicate a few of the highlights:

(*a*) Improving technology in the production and application of solvents is constantly providing a source of new and economical products to meet specialized requirements. The effects have been felt right through the vast and sprawling field of surface coatings, both in terms of decoration and protection. Further development may well prove necessary in the light of new legislation relating to alleged pollution hazards of some familiar solvent products.

(*b*) A very considerable part of the world's rubber industry is concerned with the manufacture of tyres. A substantial proportion of every tyre is represented by carbon black, which is essential to provide the reinforcement that gives the modern tyre its durability.

(*c*) In the production of synthetic ammonia, natural gas, refinery gas and liquid hydrocarbons already account for the major share of the world's production. A further substantial expansion in nitrogenous fertilizer usage is inevitable if the world is to feed its increasing population. There is every evidence that most of this further expansion will rely upon petroleum sources of raw material. This factor has a particularly important bearing on the overall statistics of the industry, since petroleum-based ammonia is already the most substantial product in terms of tonnage in the petroleum chemical range.

(*d*) Scientific farming methods rely not only on fertilizers for the stimulation of plant growth, but also upon a wide range of organic chemical pesticides to control the growth of unwanted plants and insects. This is a study in itself (but outside the scope of the present book) in which chemicals from petroleum are playing a part of ever-increasing importance.

Chapter 18

Changing Economic Factors in Petroleum Chemical Manufacture

The reason for the development of chemicals from a petroleum source as an alternative to, or at the expense of, other sources, is an economic one. Hydrocarbons have formed a readily available, economical, and flexible raw material for almost all organic chemical manufacture, and for certain inorganic chemical manufactures.

The core of the petroleum chemical industry is represented by the production of the lower olefins. As has been pointed out, the U.S.A. has found it most economical to develop its ethylene production mostly upon ethane extracted from 'wet' natural gas. In passing, it may be observed that where ethylene is the only olefin needed on a given site, then ethane cracking has been the most economical way of getting it.

Europe, having no suitable sources of ethane, turned, after a hesitant start, to naphtha cracking, by means of which they produced ethylene, propylene, a C_4 stream and a source of aromatics in one operation. As the values of some of the co-products were more readily appreciated, naphtha cracking was seen to be a rather desirable operation for the production of these materials. As the long term availability of ethane for cracking to ethylene in the U.S.A. becomes more questionable, there has been an increasing effort to persuade the U.S. Government to allow unlimited naphtha imports for cracking to olefins. Some remarkable differences of opinion on the relative advantages of these feedstocks have arisen. The truth of the matter would seem to be that with such an elaborate complex of product values to consider, every individual company will reach a different answer. Even though it may be decided as a matter of economic policy that naphtha imports to the U.S.A. should be permitted on an increased scale, it is by no means certain that the required quantities could be made available, without creating a supply problem which would affect the price and thereby defeat the objective. This position may be accentuated by a substantial demand in the U.S.A. for naphtha as a feedstock for 'synthetic natural gas'.

It would seem inevitable that in such areas where naphtha cracking is not already the dominating source of ethylene, there will be an increasing tendency towards the use of liquid feedstocks (naphtha or gas oil) in the future. This will tend to intensify the effect already observable, whereby the production of petroleum chemicals is concentrated in huge clusters of plants at a limited number of manufacturing centres.

Where naphtha cracking is the major source of ethylene, it is apparent that a major differential in the rate of development of the co-products compared with ethylene could cause problems. This could arise in Europe, for example, with butadiene, and possibly in the long run, propylene. If the growth of butadiene usage does not match that of ethylene, there could develop a growing surplus of butadiene. In the past this has led to the shutting down of dehydrogenation plants for butadiene manufacture in Europe, and in the future the position may be alleviated by export possibilities.

The main prospect of adjusting the balance of production is the development in Europe of 'natural gas liquids' as ethylene feedstock. One possibility is that such materials might become available from the North Sea. Alternatively, it might be more economical to import ethane and propane from outside Europe, rather than to continue to extend naphtha cracking without any chemical value for the butadiene produced.

We face the somewhat ironical situation that Europe could be moving towards a mixed feedstock pattern for ethylene by introducing natural gas liquids, while the U.S.A. is moving towards the same pattern from the opposite direction—they are becoming increasingly short of natural gas liquids, and this is turning their attention towards the liquid hydrocarbon feedstocks.

One of the most noteworthy factors in the manufacture of chemicals from petroleum has been the very considerable increase in the scale of individual production units in the past fifteen years. Reference to the 'benefits of scale' was made early in Chapter 2, and the simple fact is that an increase of capacity at the design stage of a plant can be achieved without a corresponding increase in cost.

For many years of exciting development the combination of technical innovation and increasing scale of production enabled the producers of these chemicals to reduce their prices in the face of an inflationary world.

In the early years of petroleum chemical manufacture the economic pattern was one not unfamiliar in the chemical industry. The scale of production was modest, and the unit margin of profit relatively high. If, therefore, you built your plant too small it was still possible to buy the balance from another producer and make a merchanting profit.

At this early stage also the design of these plants was relatively unsophisticated. This usually meant excessive safety factors in the design calculation. It was normal for the rated capacity to be considerably exceeded once experience had been gained with the operation of the plant. This 'extra' capacity was a source of valuable and profitable additional revenue.

As the pace of growth accelerated the increased tonnage was offset by smaller profit margins. The economic penalty for building a plant too small was sharply accentuated. The pressures to build the plant to maximum size were growing all the time. As the prices came down the market growth was stimulated further.

This cycle of expansion and opportunity continued without a significant break until the later 1960's. At this point a number of unfavourable factors took simultaneous effect.

The benefits of scale are particularly evident at the lower end of the scale. A growth from, say, 10 000 tons to 100 000 tons per year offers enormous economic benefits. The next stage of growth from 100 000 tons to 200 000 tons per year, which may be equally difficult to achieve, offers far less benefit. On the other hand, the risks of a malfunctioning or delay in the plant are almost directly proportional to the scale. As the plants became bigger and bigger the cost of a failure (or partial failure) grew at a much faster rate than the benefit of a success. A few well-publicized problem plants brought the situation home to a number of manufacturers.

The period 1970–71 saw an unprecedented jump in capital costs for chemical plant. During a two-year period rises of perhaps 20–30 per cent were recorded, with every indication of a continuing inflationary trend. For much of the period 1950–70 petroleum chemical production had benefited from its dependence on capital rather than the more volatile labour costs. Suddenly the position went into reverse. Most of the scale benefits had already been absorbed, and suddenly capital and labour costs jumped ahead. There was every sign also that two decades of low oil prices were coming to an end.

Other economic factors, in so far as they were changing, were tending in an adverse direction. In general, notably in the United Kingdom, petroleum chemical producers would claim to have followed enlightened policies of dealing with gaseous and liquid effluents, and the many factors that come into the now fashionable concern with the environment. It seems clear that the future will introduce ever more stringent requirements in this respect.

In the field of innovation it may be commented as a generalization that the costs are increasing, and the benefits are harder to achieve. The costs, for example, of conducting research, or introducing a new pesticide or a new drug, continually increase. The scope for exploiting a major new polymer is clearly much limited by the group of existing economical products in large-scale production and covering a wide range of properties.

The sum total of these effects was that by 1970 petroleum chemical producers found that their profitability was eroded. Their customers had for many years assumed with some justification that prices were on an inexorable downwards trend. In many instances, due to an untimely business lull which occurred just at the peak growth of cost, there was spare capacity. To cope with the substantially increased costs the producers naturally wished to raise prices. This was resisted by customers who regarded reducing prices as some kind of natural law in this business. So long as over-capacity persisted, consumers could play off producers against each other.

The inevitable effect of this was a marked slowing down in forward investment in the industry. Such factors can be slow to take effect, but unless the current cycle of events can be broken, one would expect the investment slowdown to cause a stringency in supplies of some basic materials by 1973–74. At this point a price increase would become inevitable to justify investment in further production, but there could be deficits in availability of some products until the newly stimulated investment could take effect in the form of manufacturing facilities.

It would be foolish to make detailed predictions about the future of this fascinating industry, but some reassessment of the scope for expansion is already taking place. The fundamental need for petroleum chemical products on an expanding scale is scarcely in doubt. It could be that a period of economic consolidation may be called for before confidence is fully restored, and this might lead to rates of growth somewhat lower than the remarkable levels achieved in the 1960's.

Chapter 19

Statistics of Petroleum Chemical Production

In presenting a brief statistical outline of petroleum chemical manufacture in various areas, it is perhaps appropriate to add a few words of caution. Such statistics commonly look very authoritative, but they contain a few traps for the unwary. Factors which may complicate the interpretation include: incomplete production figures (certain statistics may exclude production for military purposes, for example), inclusion of extraneous items (for instance, plastics statistics may relate to products as sold, including plasticizers, fillers, etc.), the difficulty of obtaining authoritative figures for materials produced largely for captive use (i.e. for further synthesis on the same site), and the problem of double counting. In the case of petroleum chemicals, one also has to contend with the basic difficulty of precise definition.

It is also relevant here to mention the pitfalls offered by the various units in which statistics are presented. In the United Kingdom we are accustomed to production statistics in long tons, and imperial gallons. Statistics from the U.S.A. are more often quoted in short tons (i.e. of 2 000 lb each) and U.S. gallons (approximately five-sixths of an imperial gallon). Continental European countries use the metric ton (of 2 204 lb or 1 000 kgm). It is not uncommon for statistics to leave the precise nature of the units undefined. This book normally uses the British units, but includes others where specified. By the time the next edition appears, we may perhaps be Europeanized to the stage of quoting metric units throughout.

The seeker after minor discrepancies in statistical information will, almost certainly, find a few to shake his head over. This is inevitable (unless the author is to 'edit' the figures) in a field so large, so broadly based, so rapid in development and so complex in integration as the one which is the subject of this book. Figures of production and consumption are taken from appropriate authorities (including the U.S. Tariff Commission, U.K. Department of Trade and Industry, the estimates put forward by major journals, and presentations before scientific, technological and commercially orientated associations). They may, therefore, be treated as informed judgments, rather than as precisely factual information.

The production figures which follow illustrate the importance of petroleum chemicals within the chemical industry. In passing, we should make some reference to the significance of this production in terms of the petroleum industry. This is of some importance if one bears in mind the frequently reiterated (and so far unjustified) fears of the imminence of the depletion of this planet's petroleum resources. The net usage of petroleum products as chemical raw materials may be typified by the U.S.A. position. There, some 5–5.5 per cent of petroleum product demand is for chemical production. A substantial part of this comes from the natural gas resources of the U.S.A. The proportion varies in other parts of the world. In areas where the chemical industry is highly developed and the petroleum resources do not include much natural gas, chemical production may account for 6–7 per cent of petroleum demand. It is equally clear that where petroleum industry development is in advance of chemical development (as happens in many producing areas) the proportion used for chemicals will be very small indeed. Across the world as a whole the proportion of petroleum products used for chemicals will probably not differ greatly from the U.S. figure. This proportion is increasing, but as petroleum usage itself is growing quite rapidly, the rate at which the proportion used as chemical raw material is growing, is quite slow.

Various suggestions have been made that by the end of this century between 8 and 15 per cent of petroleum products will be consumed in chemical production. On present evidence, perhaps 11–12 per cent is the most reasonable estimate for this proportion in the year 2000. It can be seen that any contribution which chemicals may make to the depletion of the world's oil reserves is relatively slight. It could also be argued that petroleum products are better used to make something tangible and useful, as an alternative to being burned to produce energy.

PRODUCTION FIGURES FOR THE U.S.A. (in millions of long tons)

	Total chemicals produced	*Total petroleum chemicals produced*	*Petroleum chemicals %*
1935	8.9	0.5	5
1940	11.2	1.1	10
1945	29.0	4.5	15
1950	33.5	7.2	21
1955	60.2	14.4	24
1960	82.3	24.7	30
1965	129.3	42.3	33
1966	139.3	48.9	35
1968	154.8	59.5	38
1969	164.3	66.2	40
1970	170.1	69.9	41

The total production of these petroleum chemicals may be further broken down as follows:

	Percentage of U.S. petroleum chemicals		
	Aliphatic organic	*Aromatic organic*	*Inorganic*
1940	54.4	2.1	43.5
1950	72.9	5.9	21.2
1960	58.2	14.3	27.5
1965	58.3	15.8	25.9
1966	57.3	15.4	27.3
1968	56.7	17.4	25.9
1969	57.1	17.6	25.3
1970	59.5	15.9	24.6

There is scope for variability in these answers as they depend upon the stage of synthesis taken for counting, the definition of what is a petroleum chemical, and the avoidance of double counting. These figures are internally consistent, having been based on a system worked out many years ago by Union Carbide.

The essence of any such system is that each sequence of production stages must be counted systematically, but once only. Illustrating this, if we take the series ethylene (raw material) → ethylene oxide (intermediate) → ethylene glycol (finished product), the stage used for production counting is the intermediate stage (in the above example, ethylene oxide).

In view of the obscurities inevitable in these assessments, a single precise figure for the percentage of U.S. organic chemicals made from petroleum is always questionable, but a figure of about 95 per cent is the normal assumption.

For the immediate future an average growth of 8–10 per cent a year is foreseen for petroleum chemicals in the U.S.A., whereas chemicals as a whole are exhibiting an annual growth more like 6 per cent.

Petroleum Chemicals in Western Europe

Statistics for petroleum chemicals as a whole have never been generally available for western Europe. The main source of such information is from publications of the O.E.C.D. and these data are delayed and incomplete.

An approximate interpretation for the 1969 figures would give the following:

Organic chemicals from petroleum ('000 long tons)

West Germany	5 100
United Kingdom	3 245
France	2 354
Italy	2 257
Total west Europe	14 320

The proportion of organic chemical production attributable to petroleum sources is quoted for European countries as follows:

Percentage organic chemicals from petroleum

	1965	*1968*	*1969*
Austria	13	25	32
France	67	81	88
West Germany	63	86	88
Italy	88	93	95
Spain	75	60	65
Sweden	72	57	74
United Kingdom	71	87	90
Average for west Europe	71	86	88

There are some slightly anomalous figures, but the trend is unmistakable. As recently as 1960 the average proportion of western Europe's organic chemicals made from petroleum was estimated to be 37 per cent.

One should beware of comparing European tonnage figures quoted here with those for the U.S.A. The latter are counted at the intermediate stage, whereas the European production is taken at the raw material stage. Moreover, the European figures do not include petroleum-based inorganic chemicals.

The Future of Petroleum Chemicals

Forecasts of various aspects of petroleum chemicals in the future have been made on a number of occasions. Notable amongst these have been estimates by M. Spaght, R. Boulitrop, P. W. Sherwood and the centenary issue of the *Oil, Paint and Drug Reporter* (now renamed *Chemical Marketing Reporter*).

Spaght's forecasts were as follows:

Production of petroleum-based chemicals
(in million tons)

Area	*1985*	*2000*
U.S.A.	80	210
Western Europe	88	240
Rest of the world (ex. communist countries)	32	150
Total	200	600

These figures are clearly based on counting at the raw material stage. Even so, they look low in comparison with some alternative estimates. If current forecasts are to be believed, the production of world plastics will reach 100 million tons in 1980 or soon after.

Some points of general agreement emerge. The production of western Europe should become comparable with that of the U.S.A. in the period 1975–85. Looking at certain basic products, the west European ethylene annual total is generally forecast to reach around 16–18 million long tons by 1980, at which time the U.S. total is also expected to be around 18 million tons. The annual demand for benzene in western Europe will rise to about 5 million long tons by 1975, and the U.S. demand will also approximate to that figure in the same year. A similar pattern will be found for a range of important products.

The other fact that emerges is that Japan will consolidate its position as number two in the petroleum chemical field. After an extraordinary growth during the 1960's, Japan had reached a stage by 1970 in which her production of petroleum chemicals was approaching half the total for western Europe.

Long term projections to the year 2000 for certain U.S. petroleum chemicals have put forward the following estimates:

U.S. production of major chemicals in year 2000

Ethylene	53.7 million long tons
Propylene	18.8 million long tons
Butadiene	5.4 million long tons
Benzene	18.8 million long tons
Toluene	14.6 million long tons
Xylenes	7.2 million long tons

World requirements for 1980 have been estimated by R. Boulitrop as follows:

Ethylene	49.2 million long tons
Propylene	33.6 million long tons
Butadiene	5.1 million long tons
Benzene	18.5 million long tons
Xylenes	5.5 million long tons
Ammonia	104.0 million long tons

So far as organic chemicals are concerned, the proportion now made worldwide from petroleum must be about 90 per cent, compared with 80 per cent in 1965. Spaght has forecast that this proportion will rise to 98 per cent by 1985 and to 99 per cent by the turn of the century.

General Comments

Statistically the production of inorganic chemicals from petroleum is not well documented, particularly in Europe. The most significant items of inorganic chemical production where petroleum-based production is important, are ammonia, sulphur and carbon black. The production of inorganic chemicals from petroleum in western Europe must have reached 11–12 million long tons in 1970. This compares with about 1.2 million long tons in 1958 and 250 000 long tons in 1954. Petroleum raw materials are far less significant in inorganic chemical production than in the case of organic chemicals. Inorganic chemicals have not grown as fast as organic chemicals in the past twenty-five years.

For the United Kingdom the peak years for growth were in the decade 1954–64 when there was an annual growth rate for petroleum chemicals in excess of 20 per cent. More recently this growth rate has subsided to a relatively sober 10–15 per cent per year.

A similar picture appears for continental Europe as a whole. In the later 1960's growth in petroleum chemicals maintained an annual rate of around 20 per cent. Here, too, the forecast for the future is in the range 12–15 per cent per year.

Some of the most astonishing figures of all came from Japan in the 1960's. Figures for growth percentages sound almost unbelievable, but they stem from a very small base in 1960. Even here there are clear signs of a more mature and sober pattern of growth being exhibited in the future.

It is obvious that, as the scale of the petroleum chemical industry grows, the maintenance of a constant percentage rate of further growth becomes more and more difficult. The more mature and well-established the industry, the less scope there is for spectacular growth. A further point that has stimulated growth in petroleum chemicals in the past has been the replacement of alternative materials (e.g. coal chemicals, fermentation chemicals, fats and oils, etc.). Once the entire range of organic chemical production has become dominated by petroleum raw materials, this type of expansion by replacement is no longer possible.

In many countries it has been noted that the rate of growth of the chemical industry is about twice the rate of growth of the economy as a whole. This is approximately true of the United Kingdom, where a chemical industry growth averaging 6 per cent per year over an extended period compares with the overall growth in the national economy which is struggling to reach 3 per cent per year. In establishing this widespread differential in favour of the chemical industry, the outstanding growth in the production of chemicals from petroleum has played a major part.

Bibliographical Note

Books on petroleum chemicals have not been numerous. Some relevant recent titles include:

Goldstein and Waddams, *The Petroleum Chemicals Industry* (3rd ed., 1967). E. & F. N. Spon, London.
Hahn, *The Petrochemical Industry* (1970). McGraw Hill, New York.
Leprince, Chauvel and Catry, *Procédés de Pétrochimie* (1971). Editions Technip, Paris.
Stobaugh, *Petrochemical Manufacturing and Marketing Guide*. Gulf Publishing Co., Houston. Vol. 1 (1967), *Aromatics and Derivatives*. Vol. 2 (1968), *Olefins, Diolefins and Acetylene*.
The Shrinking World of Petrochemicals, Toronto conference, May 1970. American Chemical Society (Division of Chemical Marketing & Economics), New York.
Petrochemicals and their Raw Materials in Europe, Budapest conference, October 1970. European Chemical Marketing Research Association, London. (Published by Technoinform, Budapest.)

Certain products have been dealt with comprehensively. These include:

Miller, *Acetylene* (Vol. 1, 1965. Vol. 2, 1966). Ernest Benn Ltd, London.
Miller (ed.), *Ethylene* (1969). Ernest Benn Ltd, London.
Monick, *Alcohols* (1968). Reinhold, New York.

It could be mentioned that the series represented by *Ethylene* and *Acetylene* is to be supplemented by books on *Propylene* and *Benzene* during the course of 1973.

Books of a more general nature include the following:

Faith, Keyes and Clark, *Industrial Chemicals* (3rd ed., 1965). Chapman and Hall, London.
This is a treasury of information on a wide range of industrial chemicals, almost exclusively limited to U.S. conditions. If previous chronology is followed, another edition is due in 1972–73.
Stephenson, *Introduction to the Chemical Process Industries* (1966). Reinhold, New York.
A useful survey of the activities of the U.S. chemical industry.

The next category to be considered is the group of encyclopaedias covering related fields:

Kirk-Othmer, *Encyclopedia of Chemical Technology*. Interscience, New York.
Encyclopedia of Polymer Science and Technology. Interscience, New York.
Chemical Economics Handbook. Stanford Research Institute, Menlo Park, California.
Modern Plastics Encyclopedia. Issued annually by Modern Plastics (McGraw Hill).

Dealing with some industrial activities in which petroleum chemicals play a prominent part are the following books:

Moncrieff, *Man-Made Fibres* (5th ed., 1970). Butterworth, London.
Duck, *Polymers-Plastics and Rubbers* (1971). Butterworth, London.
Gait and Hancock, *Plastics and Synthetic Rubbers* (1970). Pergamon Press, Oxford.
Brydson, *Plastics Materials* (2nd ed., 1969). Butterworth, London.
Nylen and Sunderland, *Modern Surface Coatings* (1965). John Wiley, New York.
Slack, *Chemistry and Technology of Fertilizers* (1967). Interscience, New York.

A constant source of new information on petroleum chemicals is to be found in a number of chemical journals. Prominent amongst those published in English are *Chemistry and Industry*, *Chemical Age*, *European Chemical News* (published in the United Kingdom); *Chemical Week*, *Chemical and Engineering News*, *Chemical Engineering*, *Chemical Engineering Progress*, *OPD Chemical Marketing Reporter* (published in the U.S.A.); the International Edition of *Chemische Industrie* (published in Germany); *Chemical Economy and Engineering Review*, *Japan Chemical Week* (published in Japan); and *Canadian Chemical Processing* (published in Canada).

Such journals include, *inter alia*, news of new developments in the petroleum chemical field, as well as feature articles covering various aspects of the subject. Much of the statistical information included in the text has been drawn from such sources.

Petroleum journals also display a lively interest in their industry's association with chemical manufacture. Notable amongst these are *Oil and Gas Journal* with the associated *Petroleum and Petrochemical International*, *Hydrocarbon News* and *Hydrocarbon Processing*. The last-named makes a practice of publishing every two years a handbook of petroleum chemical processes detailing relevant process, economic and licensing data and including flow sheets, which form the basis of those presented in this book.

Some journals serving other process industries which have a link with petroleum chemicals include *Plastics and Rubber Weekly*, *British Plastics*, *Modern Plastics International*, *Plastics Industry News* (Japan), *Polymer Paint and Colour Journal*, *Rubber World*, *Rubber Age*, *Rubber Statistical Bulletin*, *Soap Cosmetics and Chemical Specialities*, *Soap Perfumery and Cosmetics*, *Nitrogen*, *Textile Organon*.

Among other reference works are the annual issues of *The Chemical Industry* by the Organization for Economic Co-operation and Development.

A source of much information of a commercial nature stems from the conferences and symposia of bodies such as the Chemical Marketing & Economics Division of the American Chemical Society, the Chemical Marketing Research Association (of the U.S.A.) and its counterpart, the European Chemical Marketing Research Association. Proceedings of these conferences are frequently published in book form (and two examples are quoted in the first section of this note).

For those who find the spate of information too much to cope with, reference might be made to the various service companies who produce prepared and indexed digests of information on specific sections of the subject. Such services are offered by Search, and Predicasts in the U.S.A., or Temple Press Data in the United Kingdom.

Additional useful points of reference are the series of books and pamphlets issued, for example, by the Noyes Development Corporation and by Charles Kline and Co.

In a somewhat different category are the major product surveys offered by a whole range of consultant companies. Notable in this group are the studies of basic products such as olefins and aromatics offered by the Pace Company of Houston and by Parpinelli of Milan.

Major reactions in chart form

(1) Acetylene
(2) Ethylene
(3) Propylene

Some Important Reactions of Acetylene

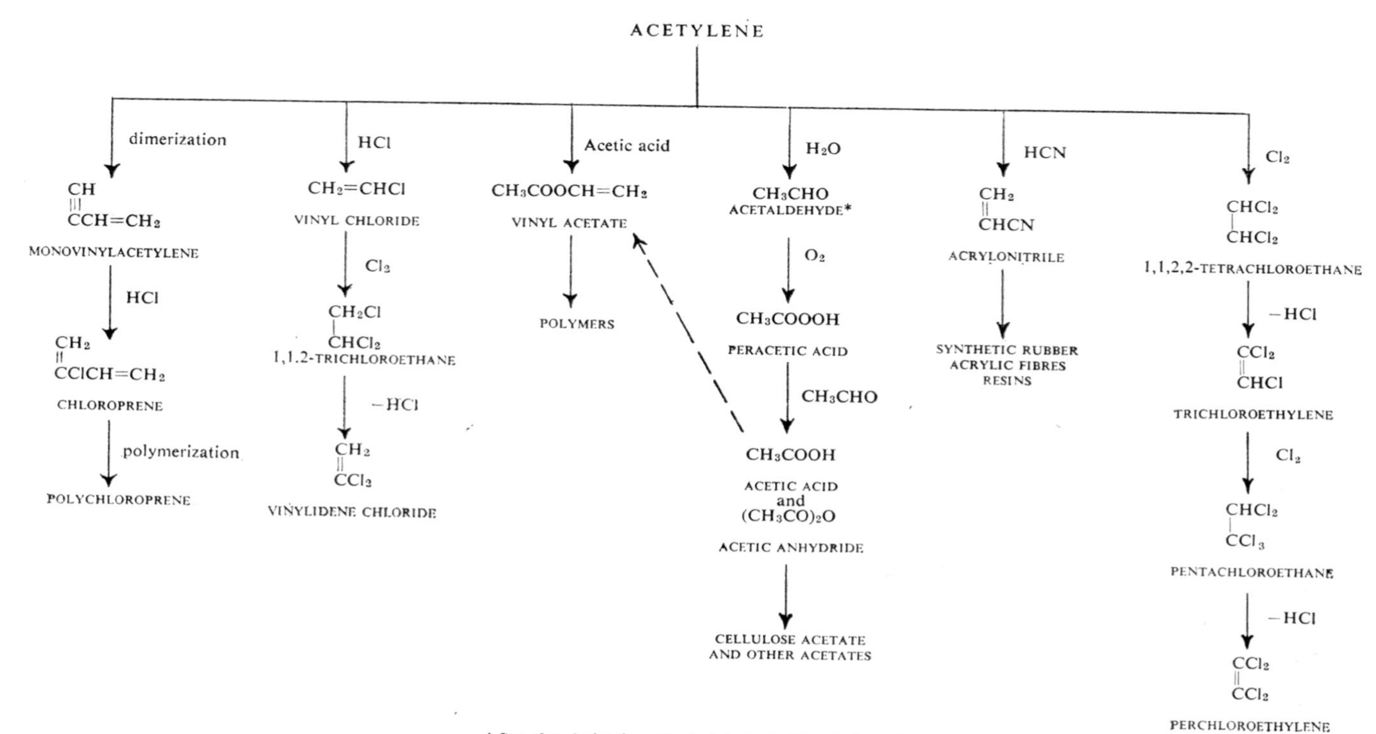

* See also derivation of n-butyl alcohol in ethylene chart.

Some Important Reactions of Ethylene

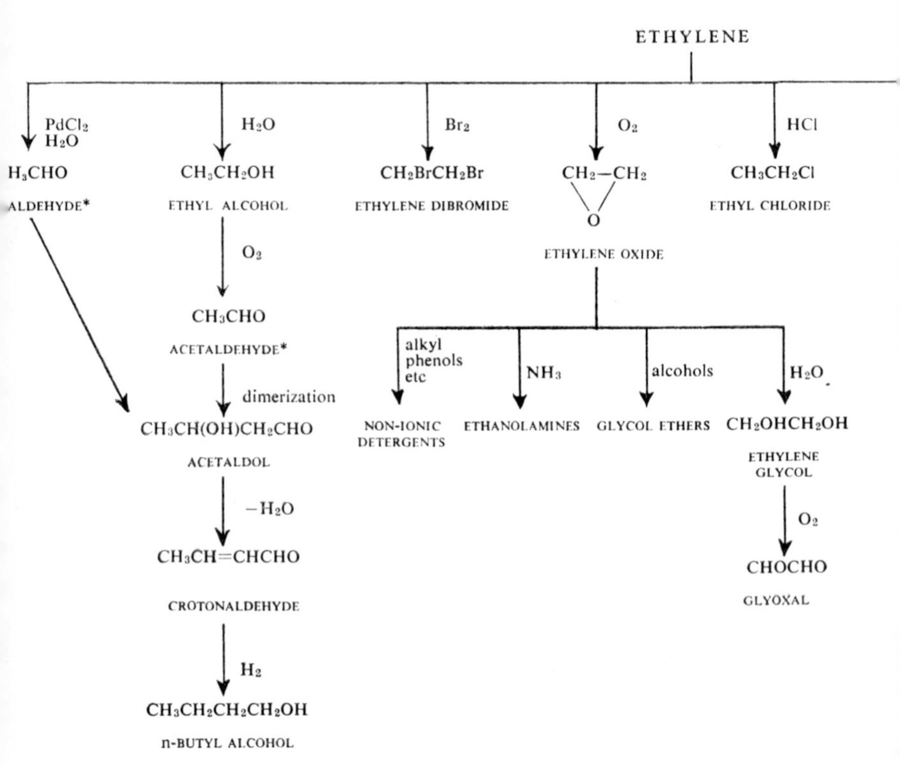

* See also derivation of acetic acid, acetic anhydride and acetates in acetylene chart.

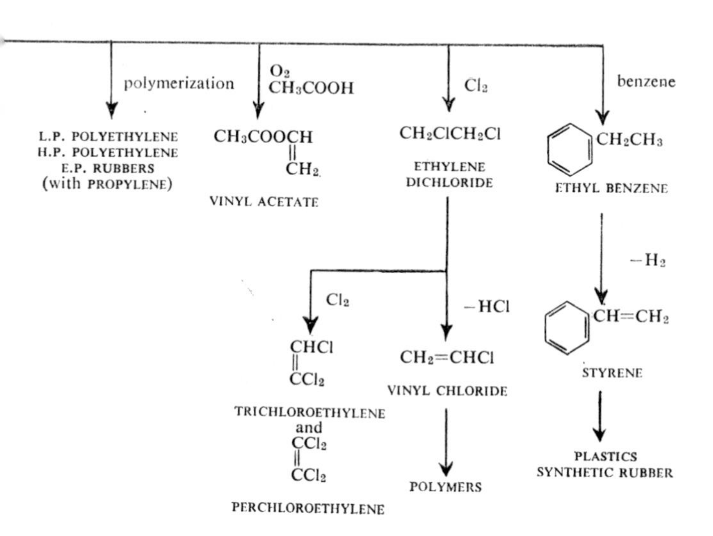

polymerization
O_2
CH_3COOH
Cl_2
benzene
L.P. POLYETHYLENE
H.P. POLYETHYLENE
E.P. RUBBERS
(with PROPYLENE)
$CH_3COOCH=CH_2$
VINYL ACETATE
CH_2ClCH_2Cl
ETHYLENE
DICHLORIDE
$C_6H_5CH_2CH_3$
ETHYL BENZENE
$-H_2$
Cl_2
$-HCl$
$CHCl=CCl_2$
TRICHLOROETHYLENE
and
$CCl_2=CCl_2$
PERCHLOROETHYLENE
$CH_2=CHCl$
VINYL CHLORIDE
POLYMERS
$C_6H_5CH=CH_2$
STYRENE
PLASTICS
SYNTHETIC RUBBER

Some Important Reactions of Propylene

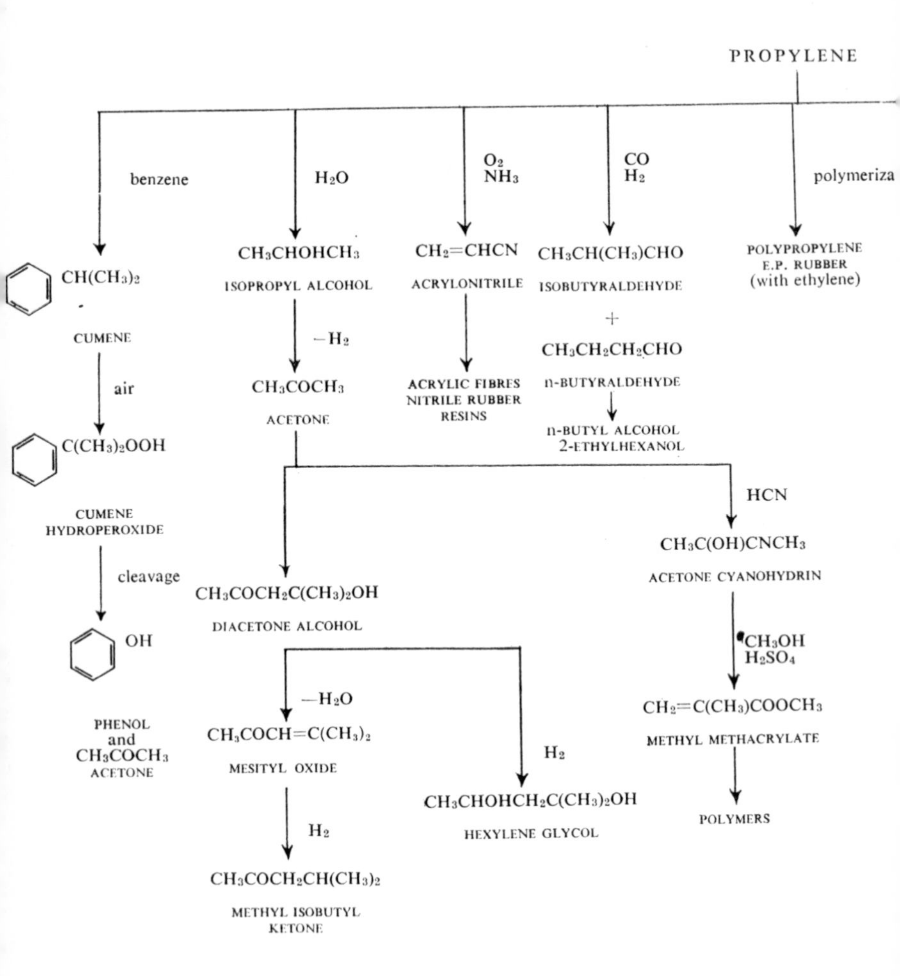

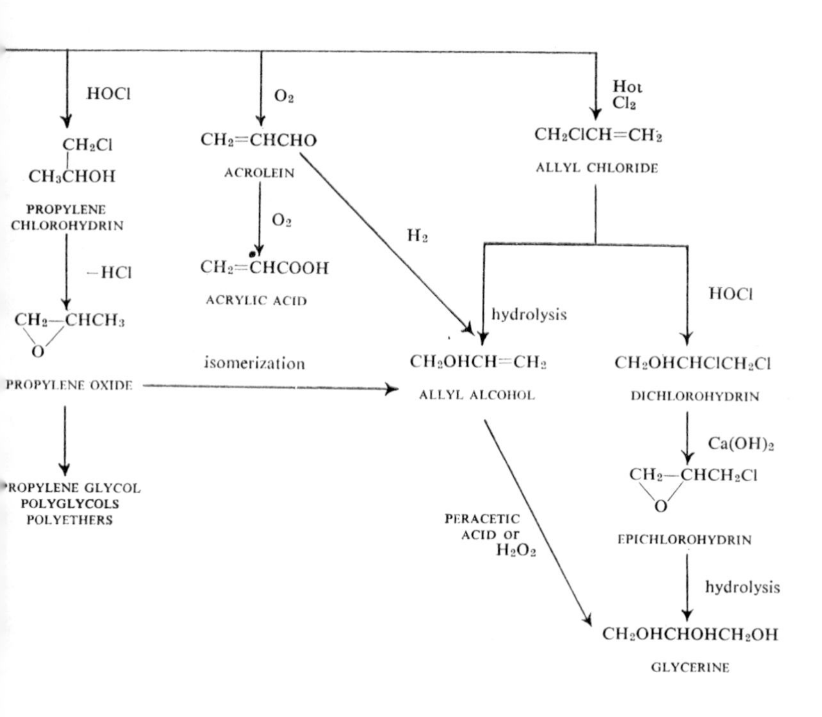
HOCl
CH_2Cl
CH_3CHOH
PROPYLENE CHLOROHYDRIN
−HCl
$CH_2—CHCH_3$
O
PROPYLENE OXIDE
PROPYLENE GLYCOL
POLYGLYCOLS
POLYETHERS
O_2
$CH_2=CHCHO$
ACROLEIN
O_2
$CH_2=CHCOOH$
ACRYLIC ACID
isomerization
H_2
Hot
Cl_2
$CH_2ClCH=CH_2$
ALLYL CHLORIDE
hydrolysis
HOCl
$CH_2OHCH=CH_2$
ALLYL ALCOHOL
$CH_2OHCHClCH_2Cl$
DICHLOROHYDRIN
$Ca(OH)_2$
$CH_2—CHCH_2Cl$
O
EPICHLOROHYDRIN
PERACETIC ACID or H_2O_2
hydrolysis
$CH_2OHCHOHCH_2OH$
GLYCERINE

Index